Compact Textbooks in Mathematics

This textbook series presents concise introductions to current topics in mathematics and mainly addresses advanced undergraduates and master students. The concept is to offer small books covering subject matter equivalent to 2- or 3-hour lectures or seminars which are also suitable for self-study. The books provide students and teachers with new perspectives and novel approaches. They may feature examples and exercises to illustrate key concepts and applications of the theoretical contents. The series also includes textbooks specifically speaking to the needs of students from other disciplines such as physics, computer science, engineering, life sciences, finance.

- **compact:** small books presenting the relevant knowledge
- **learning made easy:** examples and exercises illustrate the application of the contents
- **useful for lecturers:** each title can serve as basis and guideline for a semester course/lecture/seminar of 2-3 hours per week.

Ana Cannas da Silva • Özlem Imamoğlu •
Alessandra Iozzi

Tensors for Scientists

A Mathematical Introduction

Ana Cannas da Silva
Department of Mathematics
ETH Zurich
Zurich, Switzerland

Özlem Imamoğlu
Department of Mathematics
ETH Zurich
Zurich, Switzerland

Alessandra Iozzi
Department of Mathematics
ETH Zurich
Zurich, Switzerland

ISSN 2296-4568   ISSN 2296-455X (electronic)
Compact Textbooks in Mathematics
ISBN 978-3-031-94135-1   ISBN 978-3-031-94136-8 (eBook)
https://doi.org/10.1007/978-3-031-94136-8

Mathematics Subject Classification: 15-xx, 15A72, 53A45, 15A69, 46A32, 47A80, 46B28, 46M05, 62J10

© The Editor(s) (if applicable) and The Author(s), under exclusive license to Springer Nature Switzerland AG 2025

This work is subject to copyright. All rights are solely and exclusively licensed by the Publisher, whether the whole or part of the material is concerned, specifically the rights of reprinting, reuse of illustrations, recitation, broadcasting, reproduction on microfilms or in any other physical way, and transmission or information storage and retrieval, electronic adaptation, computer software, or by similar or dissimilar methodology now known or hereafter developed.
The use of general descriptive names, registered names, trademarks, service marks, etc. in this publication does not imply, even in the absence of a specific statement, that such names are exempt from the relevant protective laws and regulations and therefore free for general use.
The publisher, the authors and the editors are safe to assume that the advice and information in this book are believed to be true and accurate at the date of publication. Neither the publisher nor the authors or the editors give a warranty, expressed or implied, with respect to the material contained herein or for any errors or omissions that may have been made. The publisher remains neutral with regard to jurisdictional claims in published maps and institutional affiliations.

This book is published under the imprint Birkhäuser, www.birkhauser-science.com by the registered company Springer Nature Switzerland AG
The registered company address is: Gewerbestrasse 11, 6330 Cham, Switzerland

If disposing of this product, please recycle the paper.

Preface

This text deals with physical or geometric entities, known as **tensors**, which can be thought of as a generalization of vectors. Tensors are central in Engineering and Physics because they provide the framework for formulating and solving problems in areas such as Mechanics (inertia tensor, stress tensor, elasticity tensor, etc.), Electrodynamics (electrical conductivity and electrical resistivity tensors, electromagnetic tensor, magnetic susceptibility tensor, etc.), or General Relativity (stress energy tensor, curvature tensor, etc.). Thus tensors are part of most Engineering and Physics curricula.

The Goal of This Text

Our aim is to provide a bridge between Linear Algebra and Multilinear Algebra, thus introducing tensors with just elementary knowledge of Linear Algebra. In particular, we focus on finite-dimensional vector spaces. After this exposure, the interested reader might want to seek other texts, such as those by Dodson [5], Jeevanjee [8], Landsberg [9], or Simmonds [11], for a solid understanding of the mathematical nature of tensors.

What Type of Text This Is

This text fits in style between a textbook and lecture notes. We hope that it will serve as an introduction or as a road map, providing just enough material to satisfy basic needs for the practical use of tensors, or to serve as a jumping board for more substantial endeavors, in Mathematics, or Engineering, or Physics.

How This Text Came to Be

This text originated as lecture notes for the course *Multilinear Algebra*, which the three of us taught at ETH Zurich between 2008 and 2021 for the Bachelor's degree program in Material Sciences. Having started out as a seminar, this course was shaped by Özlem Imamoğlu (whose career started out with a Bachelor in Electrical

Engineering) in 2008 to consist of one two-hour lecture a week for 14 weeks, complemented by a one-hour weekly exercise class. The target students previously had had only a one-semester course of Linear Algebra, which had not emphasized change of coordinates' formulas. This led to a first chapter, where we give a review of Linear Algebra, with an eye toward tensors. Over the years, the three of us, sharing our materials and benefiting from the input from students and assistants, developed the contents until we arrived at the present version.

Whom This Text Is Meant For

As in the original course, the target audience remains a Science or Engineering college student with a typical one-semester course in Linear Algebra. The goal remains to provide foundations for a solid handling of the mathematical side of tensors.

Background Expected for This Text

The main prerequisite for the first four chapters is a standard one-semester first-year course in Linear Algebra, for example at the level of the books by Axler [2], Bretscher [3], or Strang [13]. That should include knowledge of eigenvalues and eigenvectors. We use the Spectral Theorem (Theorem 4.9), but do not expect it to have been covered earlier.

Some familiarity with Physics is required for the last chapter handling physical tensors. At ETH Zurich, where this text originated, the first year of a Bachelor's program covers the foundations of Physics and Mathematical Analysis. Since we only address Mechanics, probably a shorter, one-semester Physics course should do. However, those are only examples and the mathematical grasp of tensors can be achieved already with the first four chapters.

Regarding the Contents

- In Chap. 1, we give a brief general introduction to this text.
- In Chap. 2, we collect and recall definitions and key facts from Linear Algebra that play a significant role in Multilinear Algebra.
- In Chap. 3, we delve into linear, bilinear, and multilinear maps, introducing the notion of dual space.
- In Chap. 4, we address inner products and how the presence of an inner product provides identifications concealing the distinction between covariance and contravariance.
- In Chap. 5, we discuss general tensors, their different types, possible symmetries, and tensor product.
- In Chap. 6, we bridge toward concrete applications from Mechanics.

Throughout, we pay particular attention to *covariance* vs. *contravariance* issues.

The exposition is complemented with exercises, to help gain comfort with the material. We provide solutions at the end, though of course these should be avoided as much as possible for better learning results.

We Are Happy to Acknowledge

For their valuable comments and pleasant cooperation, we would like to thank our students and assistants throughout the years, in particular, Horace Chaix, Sebastian Gassenmeier, Jonas Jermann, and Dario Stocco. We are also grateful to our colleagues Alexander Caspar, Lorenz Halbeisen and José Natário for their encouragement and for corrections to an earlier version.

Zurich, Switzerland
Ana Cannas da Silva
Özlem Imamoğlu
Alessandra Iozzi

Declarations

Competing Interests The authors have no competing interests to declare that are relevant to the content of this manuscript.

Contents

1	**Introduction**			1
2	**Review of Linear Algebra**			7
	2.1	Vector Spaces		7
		2.1.1	Vectors and Scalars	7
		2.1.2	Subspaces	11
	2.2	Bases		12
		2.2.1	Definition of Basis	12
		2.2.2	Facts About Bases	15
	2.3	The Einstein Convention		18
		2.3.1	A Convenient Summation Convention	18
		2.3.2	Change of Basis	20
		2.3.3	The Kronecker Delta Symbol	23
	2.4	Linear Transformations		25
		2.4.1	Linear Transformations as (1, 1)-Tensors	25
		2.4.2	Conjugate Matrices	30
		2.4.3	Eigenbases	32
3	**Multilinear Forms**			37
	3.1	Linear Forms		37
		3.1.1	Definition and Examples	37
		3.1.2	Dual Space and Dual Basis	39
		3.1.3	Covariance of Linear Forms	42
		3.1.4	Contravariance of Dual Bases	45
	3.2	Bilinear Forms		47
		3.2.1	Definition and Examples	47
		3.2.2	Tensor Product of Two Linear Forms on V	49
		3.2.3	A Basis for Bilinear Forms	50
		3.2.4	Covariance of Bilinear Forms	53
	3.3	Multilinear Forms		54
		3.3.1	Definition, Basis and Covariance	54
		3.3.2	Examples of Multilinear Forms	55
		3.3.3	Tensor Product of Multilinear Forms	57

4 Inner Products 59
4.1 Definitions and First Properties 59
4.1.1 Inner Products and Their Related Notions 59
4.1.2 Symmetric Matrices and Quadratic Forms 61
4.1.3 Inner Products vs. Symmetric Positive Definite Matrices 63
4.1.4 Orthonormal Bases 64
4.2 Reciprocal Basis 70
4.2.1 Definition and Examples 70
4.2.2 Properties of Reciprocal Bases 73
4.2.3 Change of Basis from a Basis $\mathcal{B}$ to Its Reciprocal Basis $\mathcal{B}^g$ 76
4.2.4 Isomorphisms Between a Vector Space and Its Dual 79
4.2.5 Geometric Interpretation 81
4.3 Relevance of Covariance and Contravariance 81
4.3.1 Physical Relevance 81
4.3.2 Starting Point 83
4.3.3 Distinction Vanishes when Restricting to Orthonormal Bases 85

5 Tensors 87
5.1 Towards General Tensors 87
5.1.1 Canonical Isomorphism Between V and $(V^*)^*$ 87
5.1.2 $(2, 0)$-Tensors 89
5.1.3 Tensor Product of Two Linear Forms on V^* 90
5.1.4 Contravariance of $(2, 0)$-Tensors 91
5.2 Tensors of Type (p, q) 92
5.3 Symmetric and Antisymmetric Tensors 94
5.4 Tensor Product 96
5.4.1 Tensor Product of Tensors 96
5.4.2 Tensor Product for Vector Spaces 97

6 Some Physical Tensors 103
6.1 Inertia Tensor 103
6.1.1 Physical Preliminaries 103
6.1.2 Moments of Inertia 107
6.1.3 Moment of Inertia About any Axis 110
6.1.4 Angular Momentum 112
6.1.5 Principal Moments of Inertia 114
6.2 Stress Tensor 117
6.2.1 Physical Preliminaries 117
6.2.2 Principal Stresses 121
6.2.3 Special Forms of the Stress Tensor 123
6.2.4 Stress Invariants 125
6.2.5 Decomposition of the Stress Tensor 126
6.3 Strain Tensor 128
6.3.1 Physical Preliminaries 128

		6.3.2	The Antisymmetric Case: Rotation	130
		6.3.3	The Symmetric Case: Strain	131
		6.3.4	Special Forms of the Strain Tensor	133
	6.4	Elasticity Tensor		133
	6.5	Conductivity Tensor		135
		6.5.1	Electrical Conductivity	135
		6.5.2	Heat Conductivity	137

7 Solutions to Exercises 139

Bibliography 173

Index 175

List of Notations

$\mathbb{R}^n$	Euclidean n-space
$P_n(\mathbb{R})$	polynomials of degree n and real coefficients
$M_{m \times n}(\mathbb{R})$	real matrices of size $m \times n$
$\mathcal{E}$	standard basis of $\mathbb{R}^3$
$\begin{pmatrix} v^1 \\ \vdots \\ v^n \end{pmatrix}$	coordinates of the vector v in the standard basis
$\mathcal{B}, \tilde{\mathcal{B}}$	basis of a vector space
$[\,\cdot\,]_\mathcal{B}$	coordinates of the vector v in the basis $\mathcal{B}$
$L_{\mathcal{B}\tilde{\mathcal{B}}}$	change of basis from $\mathcal{B}$ to $\tilde{\mathcal{B}}$
span $\{b_1, \ldots, b_n\}$	the vector space consisting of linear combinations of $\{b_1, \ldots, b_n\}$
$\dim(V)$	the dimension of the vector space V
V^*	the dual of the vector space V
β^j	coordinate form
$\mathcal{B}^*$	the basis of V^* consisting of coordinate forms
δ^i_j	Kronecker delta symbol
$v^j b_j$	$\sum_{j=1}^n v^j b_j$
$V \times W$	the Cartesian product of the vector spaces V and W
$v \cdot w$	the dot product of the vectors v and w
$v \times w$	the cross product of the vectors v and w
$u \cdot (v \times w)$	the triple scalar product of the vectors u, v and w
p_A	the characteristic polynomial of the matrix A
E_λ	eigenspace corresponding to the eigenvector λ
$\|v\|$	norm of the vector v
$v \perp w$	orthogonal vectors v and w
$\text{proj}_w v$	orthogonal projection of the vector v in the direction of w
$\mathcal{E}^g$	reciprocal basis of a vector space with the standard basis $\mathcal{E}$ with respect to an inner product g
$\mathcal{B}^g$	reciprocal basis of a vector space with basis $\mathcal{B}$ with respect to an inner product g
$G_\mathcal{B}$	the matrix associated to the inner product g with respect to a basis $\mathcal{B}$

$\mathrm{Bil}(V \times V, \mathbb{R})$	the vector space of all bilinear forms $V \times V \to \mathbb{R}$
$\mathrm{Lin}(V, W)$	linear maps $V \to W$
$\mathrm{Lin}(V, W^*)$	the vector space of bilinear forms $\mathrm{Bil}(V \times W, \mathbb{R})$
$\alpha \otimes \beta$	the tensor product of two linear forms α and β
$V \otimes W$	the tensor product of the vector spaces V and W
$V^* \otimes V^*$	the vector space of bilinear forms on V
$\{\beta^i \otimes \beta^j\}_{i,j=1}^{\dim V}$	basis of $V^* \otimes V^*$ where β are coordinate forms
$T \otimes U$	the $k+\ell$-linear form obtained as the tensor product of the k-linear form T and the ℓ-linear form U.
$T^{i_1,\ldots,i_p}_{j_1,\ldots,j_q}$	the components $T(\beta^{i_1},\ldots,\beta^{i_p},b_{j_1},\ldots,b_{j_q})$ of the (p,q)-tensor T with respect to the basis $\{b_1,\ldots,b_n\}$ on V and to the dual basis $\{\beta^1,\ldots,\beta^n\}$ on V^*
$\mathcal{T}^p_q(V)$	the vector space of all (p,q)-tensors on V
$\mathcal{T}^0_k(V)$	the real vector space of all covariant k-tensors
$S^k V^*$	real vector space of all covariant symmetric k-tensors
$\bigwedge^k V^*$	the set of all antisymmetric covariant k-tensor
$T \otimes U$	the $(p+k, q+\ell)$-tensor obtained as tensor product of the (p,q)-tensor T and the (k,ℓ)-tensor U
$V^{\otimes p}$	$V \otimes \cdots \otimes V = \mathcal{T}^p_0(V)$
E	kinetic energy
I_{ij}	component of the inertia tensor with respect to an orthonormal basis
I_u	component of the inertia tensor with respect to an axis determined by the vector u
L	total angular momentum
B	magnetic fluid density
H	magnetic intensity
μ	scalar permeability
μ	tensor permeability
p	momentum

List of Tables

Table 3.1	Duality	46
Table 3.2	Covariance vs. contravariance	46
Table 4.1	Covariance and contravariance of vector coordinates	77
Table 5.1	Covariance and contravariance of aforementioned tensors	88
Table 5.2	Aforementioned tensors viewed within general definition	93

Introduction

Just like the main protagonists in Linear Algebra are *vectors* and *linear maps*, the main protagonists in Multilinear Algebra are *tensors* and *multilinear maps*. Tensors describe linear relations among objects in space. As in the case of vectors, the quantitative description of tensors, i.e., their description in terms of numbers, changes when we change the *frame of reference*, which is mostly just the *basis* as in Linear Algebra. Generalizing the case of vectors, tensors are represented—once a basis is chosen—by *multidimensional arrays of numbers* (Figs. 1.1, 1.2, and 1.3).

In the notation, the indices can be upper or lower. For tensors of order at least 2, some indices can be upper and some lower. The numbers in the arrays are called **components** of the tensor and give the representation of the tensor *with respect to a given basis*.

Two natural questions arise:

- Why do we need tensors?
- What are the important features of tensors?

Why Do We Need Tensors?

Scalars are not enough to describe directions, for which we need to resort to vectors. At the same time, vectors might not be enough, in that they lack the ability to "modify" vectors.

Example 1.1 We denote by **B** the magnetic flux density measured in $V \cdot s/m^2$ and by **H** the magnetizing intensity measured in A/m (the physical units here are: Volt V, second s, meter m, Ampere A, Henry H). They are related by the formula

$$\mathbf{B} = \mu \mathbf{H},$$

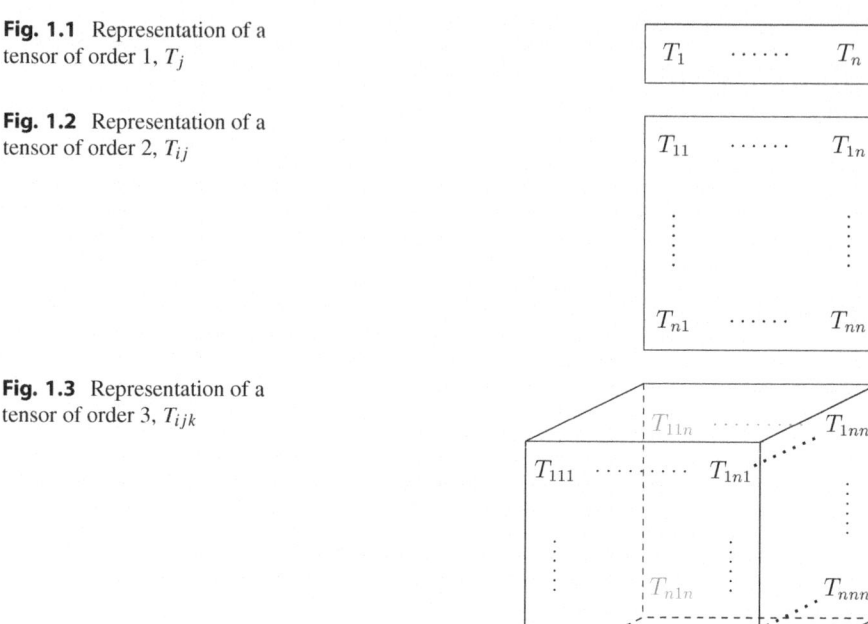

Fig. 1.1 Representation of a tensor of order 1, T_j

Fig. 1.2 Representation of a tensor of order 2, T_{ij}

Fig. 1.3 Representation of a tensor of order 3, T_{ijk}

where μ is the scalar permeability of the medium in H/m. In free space, $\mu = \mu_0 = 4\pi \times 10^{-7}$ H/m is a scalar, so that the flux density and the magnetization are vectors that differ only by their magnitude.

Some materials, however, have properties that make these terms differ both in magnitude and direction. In such materials the scalar permeability is replaced by the tensor permeability $\boldsymbol{\mu}$ and

$$\mathbf{B} = \boldsymbol{\mu} \cdot \mathbf{H}.$$

Being vectors, **B** and **H** are tensors of order 1, and $\boldsymbol{\mu}$ is a tensor of order 2. We will see that they are of different type, and in fact *the order of* **H** *"cancels out" with part of the order of* $\boldsymbol{\mu}$ *to give a tensor of order 1*. □

What Are the Important Features of Tensors?

Physical laws do not change with different coordinate systems, hence tensors describing them must satisfy some *invariance* properties. While tensors remain intrinsically invariant with respect to changes of bases, their components will vary according to two fundamental modes: **covariance** and **contravariance**, depending on whether the components change in a way parallel to the change of basis or in an opposite way.[1] We will introduce covariance and contravariance in Chap. 3 and see specific physical instances in Chap. 6.

Here are a couple of examples regarding a familiar tensor from Linear Algebra, namely a *vector*. We review the effect of a change of basis, showing that a vector is a **contravariant tensor of first order**. We use freely notions and properties that will be recalled in Chap. 2.

Example 1.2 Let $\mathcal{B} = \{b_1, b_2, b_3\}$ and $\widetilde{\mathcal{B}} = \{\tilde{b}_1, \tilde{b}_2, \tilde{b}_3\}$ be two basis of a vector space V. A vector $v \in V$ can be written as

$$v = v^1 b_1 + v^2 b_2 + v^3 b_3,$$

or

$$v = \tilde{v}^1 \tilde{b}_1 + \tilde{v}^2 \tilde{b}_2 + \tilde{v}^3 \tilde{b}_3,$$

where v^1, v^2, v^3 (resp. $\tilde{v}^1, \tilde{v}^2, \tilde{v}^3$) are the coordinates of v with respect to the basis $\mathcal{B}$ (resp. $\widetilde{\mathcal{B}}$). The reason for distinguishing lower and upper indices will become apparent already in Chap. 2.

We use the following notation:

$$[v]_{\mathcal{B}} = \begin{pmatrix} v^1 \\ v^2 \\ v^3 \end{pmatrix} \quad \text{and} \quad [v]_{\widetilde{\mathcal{B}}} = \begin{pmatrix} \tilde{v}^1 \\ \tilde{v}^2 \\ \tilde{v}^3 \end{pmatrix}, \tag{1.1}$$

and we are interested in finding the relation between the coordinates of v in the two bases.

The vectors $\tilde{b}_j$, $j = 1, 2, 3$, in the basis $\widetilde{\mathcal{B}}$ can be written as a linear combination of vectors in $\mathcal{B}$ as follows:

$$\tilde{b}_j = L_j^1 b_1 + L_j^2 b_2 + L_j^3 b_3,$$

for some $L_j^i \in \mathbb{R}$. We consider the matrix of the change of basis from $\mathcal{B}$ to $\widetilde{\mathcal{B}}$,

$$L := L_{\mathcal{B}\widetilde{\mathcal{B}}} = \begin{bmatrix} L_1^1 & L_2^1 & L_3^1 \\ L_1^2 & L_2^2 & L_3^2 \\ L_1^3 & L_2^3 & L_3^3 \end{bmatrix}$$

---

[1] The Latin prefix *co* means "joint" or "together", whereas the Latin prefix *contra* means "contrary" or "against".

whose jth-column consists of the coordinates of the vectors $\tilde{b}_j$ with respect to the basis $\mathcal{B}$. The equalities

$$\begin{cases} \tilde{b}_1 = L_1^1 b_1 + L_1^2 b_2 + L_1^3 b_3 \\ \tilde{b}_2 = L_2^1 b_1 + L_2^2 b_2 + L_2^3 b_3 \\ \tilde{b}_3 = L_3^1 b_1 + L_3^2 b_2 + L_3^3 b_3 \end{cases}$$

can simply be written as

$$(\tilde{b}_1 \ \tilde{b}_2 \ \tilde{b}_3) = (b_1 \ b_2 \ b_3) L. \tag{1.2}$$

One can check this symbolic equation using the rules of matrix multiplication. Analogously, writing basis vectors in a row and vector coordinates in a column, we can write

$$v = v^1 b_1 + v^2 b_2 + v^3 b_3 = (b_1 \ b_2 \ b_3) \begin{pmatrix} v^1 \\ v^2 \\ v^3 \end{pmatrix} \tag{1.3}$$

as well as

$$v = \tilde{v}^1 \tilde{b}_1 + \tilde{v}^2 \tilde{b}_2 + \tilde{v}^3 \tilde{b}_3 = (\tilde{b}_1 \ \tilde{b}_2 \ \tilde{b}_3) \begin{pmatrix} \tilde{v}^1 \\ \tilde{v}^2 \\ \tilde{v}^3 \end{pmatrix} = (b_1 \ b_2 \ b_3) L \begin{pmatrix} \tilde{v}^1 \\ \tilde{v}^2 \\ \tilde{v}^3 \end{pmatrix}, \tag{1.4}$$

where we used Eq. (1.2) in the last equality. Comparing the expression of v in Eqs. (1.3) and (1.4), we conclude that

$$L \begin{pmatrix} \tilde{v}^1 \\ \tilde{v}^2 \\ \tilde{v}^3 \end{pmatrix} = \begin{pmatrix} v^1 \\ v^2 \\ v^3 \end{pmatrix}$$

or, equivalently,

$$\begin{pmatrix} \tilde{v}^1 \\ \tilde{v}^2 \\ \tilde{v}^3 \end{pmatrix} = L^{-1} \begin{pmatrix} v^1 \\ v^2 \\ v^3 \end{pmatrix}.$$

We say that the components of a vector v are *contravariant* because they change by L^{-1} when the basis changes by L; see Sect. 2.3.2. A vector v is hence a *contravariant 1-tensor* or *tensor of order* $(1, 0)$. □

Example 1.3 (A Numerical Example) Let

$$\mathcal{E} = \{e_1, e_2, e_3\} = \left\{ \begin{bmatrix} 1 \\ 0 \\ 0 \end{bmatrix}, \begin{bmatrix} 0 \\ 1 \\ 0 \end{bmatrix}, \begin{bmatrix} 0 \\ 0 \\ 1 \end{bmatrix} \right\} \tag{1.5}$$

be the *standard* basis or $\mathbb{R}^3$ and let

$$\widetilde{\mathcal{B}} = \{\tilde{b}_1, \tilde{b}_2, \tilde{b}_3\} = \left\{ \begin{bmatrix} 1 \\ 2 \\ 3 \end{bmatrix}, \begin{bmatrix} 4 \\ 5 \\ 6 \end{bmatrix}, \begin{bmatrix} 7 \\ 8 \\ 0 \end{bmatrix} \right\}$$

be another basis of $\mathbb{R}^3$. The vector[2] $v = \begin{bmatrix} 1 \\ 1 \\ 1 \end{bmatrix}$ has coordinates

$$[v]_\mathcal{E} = \begin{pmatrix} 1 \\ 1 \\ 1 \end{pmatrix} \quad \text{and} \quad [v]_{\widetilde{\mathcal{B}}} = \begin{pmatrix} -\frac{1}{3} \\ \frac{1}{3} \\ 0 \end{pmatrix}.$$

Since we have that

$$\begin{cases} \tilde{b}_1 = 1 \cdot e_1 + 2 \cdot e_2 + 3 \cdot e_3 \\ \tilde{b}_2 = 4 \cdot e_1 + 5 \cdot e_2 + 6 \cdot e_3 \\ \tilde{b}_3 = 7 \cdot e_1 + 8 \cdot e_2 + 0 \cdot e_3 \end{cases},$$

the matrix of the change of coordinates from $\mathcal{E}$ to $\widetilde{\mathcal{B}}$ is

$$L = \begin{bmatrix} 1 & 4 & 7 \\ 2 & 5 & 8 \\ 3 & 6 & 0 \end{bmatrix}.$$

---

[2] For a general basis $\mathcal{B}$, the notation $[\,\cdot\,]_\mathcal{B}$ indicates the "operation" of taking the vector v and looking at its coordinates in the basis $\mathcal{B}$. However, in order to "write down explicitly" a vector (that is three real numbers that we write in column), one needs to give coordinates and the coordinates are usually given with respect to the standard basis. In this case, there is the slightly confusing fact that

$$v = \begin{bmatrix} v^1 \\ \vdots \\ v^n \end{bmatrix} \quad \text{has} \quad [v]_\mathcal{E} = \begin{pmatrix} v^1 \\ \vdots \\ v^n \end{pmatrix}.$$

Then we can check that

$$\begin{pmatrix} -\frac{1}{3} \\ \frac{1}{3} \\ 0 \end{pmatrix} = L^{-1} \begin{pmatrix} 1 \\ 1 \\ 1 \end{pmatrix}$$

or, equivalently, that

$$L \begin{pmatrix} -\frac{1}{3} \\ \frac{1}{3} \\ 0 \end{pmatrix} = \begin{pmatrix} 1 \\ 1 \\ 1 \end{pmatrix}.$$

□

Review of Linear Algebra

2

This chapter collects and recalls definitions and key facts learned in Linear Algebra, from vector spaces to linear transformations, which will play a significant role in *Multilinear* Algebra. Along the way, we fix some notations and standards for this text. We introduce the Einstein convention, which we will subsequently mostly follow.

2.1 Vector Spaces

A *vector space* (or *linear space*) is a set of objects where addition and *scaling* are defined in a way that satisfies natural requirements for such operations, namely the properties listed in the definition below.

2.1.1 Vectors and Scalars

In this text, we will only consider *real* vector spaces, that is vector spaces over $\mathbb{R}$, where the scaling is by real numbers.

> DEFINITION 2.1. A **vector space** V (over $\mathbb{R}$) is a set V equipped with two operations:
>
> - *vector addition:* $V \times V \to V$, $(v, w) \mapsto v + w$, and
> - *multiplication by a scalar:* $\mathbb{R} \times V \to V$, $(\alpha, v) \mapsto \alpha v$,
>
> satisfying the following properties:
>
> (1) (associativity) $(u + v) + w = u + (v + w)$ for every $u, v, w \in V$;

(2) (commutativity) $u + v = v + u$ for every $u, v \in V$;
(3) (existence of the zero vector) There exists $0 \in V$ such that $v + 0 = v$ for every $v \in V$;
(4) (existence of additive inverse) For every $v \in V$, there exists $w_v \in V$ such that $v + w_v = 0$. The vector w_v is denoted by $-v$.
(5) $\alpha(\beta v) = (\alpha\beta)v$ for every $\alpha, \beta \in \mathbb{R}$ and every $v \in \mathbb{R}$;
(6) $1v = v$ for every $v \in V$;
(7) $\alpha(u + v) = \alpha u + \alpha v$ for all $\alpha \in \mathbb{R}$ and $u, v \in V$;
(8) $(\alpha + \beta)v = \alpha v + \beta v$ for all $\alpha, \beta \in \mathbb{R}$ and $v \in V$.

An element of the vector space is called a **vector** and, mostly in the context of vector spaces, a real number is called a **scalar**.

Example 2.2 (Prototypical Example of a Vector Space) The Euclidean space $\mathbb{R}^n$, $n = 1, 2, 3, \ldots$, is a vector space with componentwise addition and multiplication by scalars. Vectors in $\mathbb{R}^n$ are denoted by

$$v = \begin{bmatrix} x_1 \\ \vdots \\ x_n \end{bmatrix},$$

with $x_1, \ldots, x_n \in \mathbb{R}$. Addition component-by-component translates geometrically to the *parallelogram law* for vector addition, well-known in $\mathbb{R}^2$ and $\mathbb{R}^3$. □

Examples 2.3 (Other Examples of Vector Spaces) The operations of vector addition and scalar multiplication are inferred from the context.

(1) The set of real polynomials of degree $\leq n$ is a vector space, denoted by

$$V = P_n(\mathbb{R}) := \{a_n x^n + a_{n-1} x^{n-1} + \ldots + a_1 x + a_0 : a_j \in \mathbb{R}\}$$

with the usual (degreewise) sum of polynomials and scalar multiplication.
(2) The set of real matrices of size $m \times n$,

$$V = M_{m \times n}(\mathbb{R}) := \left\{ \begin{bmatrix} a_{11} & \ldots & a_{1n} \\ \vdots & & \vdots \\ a_{m1} & \ldots & a_{mn} \end{bmatrix} : a_{ij} \in \mathbb{R} \right\}$$

with componentwise addition and scalar multiplication.
(3) The space $\{f : W \to \mathbb{R}\}$ of all real-valued functions on a vector space W. Addition of functions $f : W \to \mathbb{R}$ and $g : W \to \mathbb{R}$, and their multiplication by a scalar $\alpha \in \mathbb{R}$ are defined pointwise: $(f + g)(w) := f(w) + g(w)$ and $(\alpha f)(w) := \alpha (f(w))$ for each $w \in W$.

2.1 Vector Spaces

(4) The space of solutions of a homogeneous linear (ordinary or partial) differential equation.
(5) The cartesian product

$$V \times W := \{(v, w) : v \in V, w \in W\}$$

of two real vector spaces V and W, endowed with factorwise addition and scalar multiplication.

□

Examples 2.4 (Non-examples of Vector Spaces) The operations below are the standard ones of vectors in the plane $\mathbb{R}^2$.

(1) The upper half-plane $V := \{(x, y) \mid y \geq 0\}$ is not a vector space. The sum of two vectors in V is still a vector in V, however, if $v \in V$ has a positive second component and α is a negative real number, then $\alpha v \notin V$.
(2) The union of the two odd quadrants, $V := \{(x, y) \mid xy \geq 0\}$ is not a vector space. The product of any vector in V with a scalar $\alpha \in \mathbb{R}$ is still a vector in V, however, the sum of two vectors in V is not always a vector in V.
(3) The graph of the real function $f(x) = 2x + 3$ is not a vector space. In particular, it does not contain any zero vector.

□

EXERCISE 2.5. Are the following vector spaces?

(1) The set V of all vectors in $\mathbb{R}^3$ perpendicular to the vector $\begin{bmatrix} 1 \\ 2 \\ 3 \end{bmatrix}$.

(2) The set of invertible 2×2 matrices, that is

$$V := \left\{ \begin{bmatrix} a & b \\ c & d \end{bmatrix} : ad - bc \neq 0 \right\}.$$

(3) The set of polynomials of degree exactly n, that is

$$V := \{a_0 x^n + a_1 x^{n-1} + \cdots + a_{n-1} x + a_n : a_j \in \mathbb{R}, a_n \neq 0\}.$$

(4) The set V of 2×4 matrices with last column zero, that is

$$V := \left\{ \begin{bmatrix} a & b & c & 0 \\ d & e & f & 0 \end{bmatrix} : a, b, c, d, e, f \in \mathbb{R} \right\}.$$

(5) The set of solutions $f : \mathbb{R} \to \mathbb{R}$ of the equation $f' = 5$, that is

$$V := \{f : \mathbb{R} \to \mathbb{R} : f(x) = 5x + C, C \in \mathbb{R}\}.$$

DEFINITION 2.6. A function $T : V \to W$ between real vector spaces V and W is a **linear transformation** if it satisfies the property

$$T(\alpha v + \beta w) = \alpha T(v) + \beta T(w),$$

for all $\alpha, \beta \in \mathbb{R}$ and all $v, w \in V$.

Examples 2.7 (**Linear Transformations**)

(1) An $m \times n$ matrix,

$$A = \begin{bmatrix} a_{11} & \cdots & a_{1n} \\ \vdots & & \vdots \\ a_{m1} & \cdots & a_{mn} \end{bmatrix},$$

defines a linear transformation $T : \mathbb{R}^n \to \mathbb{R}^m$ by multiplication, $T(v) := Av$.
(2) Differentiation from the set $P_n(\mathbb{R})$ of real polynomials of degree $\leq n$ to the set $P_{n-1}(\mathbb{R})$ of real polynomials of degree $\leq n - 1$ is a linear transformation, taking a polynomial $p(x)$ to its derivative $p'(x)$.
(3) Transposition from the set $M_{m \times n}(\mathbb{R})$ of real $m \times n$ matrices to the set $M_{n \times m}(\mathbb{R})$ of real $n \times m$ matrices is a linear transformation,

$$A = \begin{bmatrix} a_{11} & \cdots & a_{1n} \\ \vdots & & \vdots \\ a_{m1} & \cdots & a_{mn} \end{bmatrix} \mapsto {}^t A = \begin{bmatrix} a_{11} & \cdots & a_{m1} \\ \vdots & & \vdots \\ a_{1n} & \cdots & a_{mn} \end{bmatrix}.$$

□

EXERCISE 2.8. Are the following linear transformations?

(1) The function $T : P_n(\mathbb{R}) \to \mathbb{R}$ evaluating a polynomial $p(x)$ of degree at most n at the point $x = 1$, i.e., $T(p(x)) := p(1)$.
(2) The function $T : \mathbb{R} \to \mathbb{R}$ defined by the formula $T(x) := 2x + 3$.
(3) The function $T : M_{n \times n}(\mathbb{R}) \to M_{n \times n}(\mathbb{R})$ taking a square $n \times n$ matrix A to its square power, $T(A) := A^2$.

EXERCISE 2.9. Show that the set of all linear transformations $T : \mathbb{R}^2 \to \mathbb{R}^3$ forms a vector space.

2.1.2 Subspaces

Naturally, once we have a vector space V, we single out its subsets that are vector spaces themselves.

> DEFINITION 2.10. A subset W of a vector space V that is itself a vector space is a **subspace**.

By reviewing the properties in the definition of vector space, we see that a subset $W \subseteq V$ is a subspace exactly when the following conditions are verified:

(1) The 0 element is in W;
(2) W is *closed under addition*, that is $v + w \in W$ for every $v, w \in W$;
(3) W is *closed under multiplication by scalars*, that is $\alpha v \in W$ for every $\alpha \in \mathbb{R}$ and every $v \in W$.

Condition (1) actually follows from (2) and (3) under the assumption that $W \neq \emptyset$ (which is always satisfied for a vector space). Yet it is often an easy way to check that a subset is not a subspace.

Recall that a **linear combination** of vectors $v_1, \ldots, v_n \in V$ is a vector of the form $\alpha_1 v_1 + \ldots + \alpha_n v_n$ for $\alpha_1, \ldots, \alpha_n \in \mathbb{R}$. With this notion, the above three conditions for a subset $W \subseteq V$ to be a subspace are equivalent to the following ones:

(1)' W is non-empty;
(2)' W is *closed under linear combinations*, that is $\alpha v + \beta w \in W$ for all $\alpha, \beta \in \mathbb{R}$ and all $v, w \in W$.

Examples 2.11 (Subspaces)

(1) The set W of 2×4 matrices with last column zero is a subspace of the vector space V of all 2×4 matrices.
(2) The line $W = \{(x, 2x) \mid x \in \mathbb{R}\}$ is a subspace of the vector space $V = \mathbb{R}^2$.
(3) The space of all differentiable real-valued functions of one real variable is a subspace of the vector space of all real-valued functions of one real variable. □

EXERCISE 2.12. Are the following subspaces?

(1) The subset of all $n \times n$ real *symmetric* matrices in the vector space $V = M_{n \times n}(\mathbb{R})$. Recall that a matrix A is **symmetric** if ${}^t A = A$.
(2) The subset of all real-valued functions of one real variable with value 0 at $x = 1$ in the vector space of all real-valued functions of one real variable.
(3) The subset of all real-valued functions of one real variable with value 1 at $x = 0$ in the vector space of all real-valued functions of one real variable.

DEFINITION 2.13. If $T : V \to W$ is a linear transformation between real vector spaces V and W, then:

- the **kernel** (or *null space*) of T is the set $\ker T := \{v \in V : T(v) = 0\}$;
- the **image** (or *range*) of T is the set $\operatorname{im} T := \{T(v) : v \in V\}$.

EXERCISE 2.14. Show that, for a linear transformation $T : V \to W$, the kernel $\ker T$ is a subspace of V and the image $\operatorname{im} T$ is a subspace of W.

2.2 Bases

The key to study and to compute in vector spaces is the concept of a *basis*, which in turn relies on the fundamental notions of *linear independence/dependence* and of *span*.

2.2.1 Definition of Basis

DEFINITION 2.15. The vectors $b_1, \ldots, b_n \in V$ are **linearly independent** if $\alpha_1 b_1 + \ldots + \alpha_n b_n = 0$ implies that $\alpha_1 = \ldots = \alpha_n = 0$. In other words, if the only linear combination of these vectors that yields the zero vector is the trivial one. We then also say that the vector set $\{b_1, \ldots, b_n\}$ is *linearly independent*.

Example 2.16 The vectors

$$\begin{bmatrix} 1 \\ 0 \\ 0 \end{bmatrix}, \begin{bmatrix} 0 \\ 1 \\ 0 \end{bmatrix}, \begin{bmatrix} 0 \\ 0 \\ 1 \end{bmatrix}$$

are linearly independent in $\mathbb{R}^3$. Indeed, we have

$$\alpha_1 \begin{bmatrix} 1 \\ 0 \\ 0 \end{bmatrix} + \alpha_2 \begin{bmatrix} 0 \\ 1 \\ 0 \end{bmatrix} + \alpha_3 \begin{bmatrix} 0 \\ 0 \\ 1 \end{bmatrix} = 0 \iff \begin{bmatrix} \alpha_1 \\ \alpha_2 \\ \alpha_3 \end{bmatrix} = \begin{bmatrix} 0 \\ 0 \\ 0 \end{bmatrix} \iff \alpha_1 = \alpha_2 = \alpha_3 = 0.$$

□

2.2 Bases

Example 2.17 The vectors

$$b_1 = \begin{bmatrix} 1 \\ 2 \\ 3 \end{bmatrix}, \quad b_2 = \begin{bmatrix} 4 \\ 5 \\ 6 \end{bmatrix}, \quad b_3 = \begin{bmatrix} 7 \\ 8 \\ 0 \end{bmatrix}$$

are linearly independent in $\mathbb{R}^3$, since we have

$$\alpha_1 b_1 + \alpha_2 b_2 + \alpha_3 b_3 = 0 \iff \begin{cases} \alpha_1 + 4\alpha_2 + 7\alpha_3 = 0 \\ 2\alpha_1 + 5\alpha_2 + 8\alpha_3 = 0 \\ 3\alpha_1 + 6\alpha_2 = 0 \end{cases}$$

$$\iff \ldots \iff \alpha_1 = \alpha_2 = \alpha_3 = 0.$$

The dots are filled in using Gauss-Jordan elimination, as in Example 2.27. □

Example 2.18 The vectors

$$b_1 = \begin{bmatrix} 1 \\ 2 \\ 3 \end{bmatrix}, \quad b_2 = \begin{bmatrix} 4 \\ 5 \\ 6 \end{bmatrix}, \quad b_3 = \begin{bmatrix} 7 \\ 8 \\ 9 \end{bmatrix}$$

are linearly *dependent* in $\mathbb{R}^3$, i.e., not linearly independent. In fact,

$$\alpha_1 b_1 + \alpha_2 b_2 + \alpha_3 b_3 = 0 \iff \begin{cases} \alpha_1 + 4\alpha_2 + 7\alpha_3 = 0 \\ 2\alpha_1 + 5\alpha_2 + 8\alpha_3 = 0 \\ 3\alpha_1 + 6\alpha_2 + 9\alpha_3 = 0 \end{cases}$$

$$\iff \ldots \iff \begin{cases} \alpha_1 = \alpha_3 \\ \alpha_2 = -2\alpha_3, \end{cases}$$

so

$$b_1 - 2b_2 + b_3 = 0$$

and b_1, b_2, b_3 are not linearly independent. For instance, we say that $b_1 = 2b_2 - b_3$ is a *non-trivial linear relation* between the vectors b_1, b_2 and b_3. □

DEFINITION 2.19. The vectors $b_1, \ldots, b_n \in V$ **span** V, if every vector $v \in V$ can be written as a linear combination $v = \alpha_1 b_1 + \ldots + \alpha_n b_n$, for some $\alpha_1, \ldots, \alpha_n \in \mathbb{R}$. We then say that the vector set $\{b_1, \ldots, b_n\}$ *spans* V and write $V = \text{span}\{b_1, \ldots, b_n\}$.

Examples 2.20 (1) The vectors $\begin{bmatrix} 1 \\ 0 \\ 0 \end{bmatrix}, \begin{bmatrix} 0 \\ 1 \\ 0 \end{bmatrix}, \begin{bmatrix} 0 \\ 0 \\ 1 \end{bmatrix}$ span $\mathbb{R}^3$.

(2) The vectors $\begin{bmatrix} 1 \\ 0 \\ 0 \end{bmatrix}, \begin{bmatrix} 0 \\ 1 \\ 0 \end{bmatrix}, \begin{bmatrix} 0 \\ 0 \\ 1 \end{bmatrix}, \begin{bmatrix} 1 \\ 1 \\ 1 \end{bmatrix}$ also span $\mathbb{R}^3$.

(3) The vectors $\begin{bmatrix} 1 \\ 0 \\ 0 \end{bmatrix}, \begin{bmatrix} 0 \\ 1 \\ 0 \end{bmatrix}$ span the xy-coordinate plane (i.e., the subspace given by the equation $z = 0$) in $\mathbb{R}^3$.

EXERCISE 2.21. Show that the set of all linear combinations of given vectors $b_1, \ldots, b_n \in V$ is a subspace of V. We denote this subspace by $\text{span}\{b_1, \ldots, b_n\}$.

DEFINITION 2.22. The vectors $b_1, \ldots, b_n \in V$ form a **basis** of V, if:

(1) they are *linearly independent* and
(2) they *span V*.

Remark 2.23 We then denote a basis as an set $\mathcal{B} := \{b_1, \ldots, b_n\}$. However it is essential to know that, despite the set notation, the order of the vectors in a basis matters for the computations.

Example 2.24 The vectors

$$e_1 := \begin{bmatrix} 1 \\ 0 \\ 0 \end{bmatrix}, \quad e_2 := \begin{bmatrix} 0 \\ 1 \\ 0 \end{bmatrix}, \quad e_3 := \begin{bmatrix} 0 \\ 0 \\ 1 \end{bmatrix}$$

form a basis of $\mathbb{R}^3$. This is called the **standard basis** of $\mathbb{R}^3$ and denoted $\mathcal{E} := \{e_1, e_2, e_3\}$. For $\mathbb{R}^n$, the *standard basis* $\mathcal{E} := \{e_1, \ldots, e_n\}$ is defined similarly.

Example 2.25 The vectors in Example 2.17 span $\mathbb{R}^3$, while the vectors in Example 2.18 do not span $\mathbb{R}^3$. To see this, we recall the following facts about bases.

2.2 Bases

2.2.2 Facts About Bases

Let V be a vector space with a *finite* basis. Then we have:

(1) All bases of V have the same number of elements. This number is called the **dimension** of V and denoted $\dim V$.
(2) If $\mathcal{B} = \{b_1, \ldots, b_n\}$ is a basis of V, there is a *unique* way of writing any $v \in V$ as a linear combination

$$v = v^1 b_1 + \ldots v^n b_n$$

of elements in $\mathcal{B}$. The numbers $v^1, \ldots, v^n$ are the **coordinates** of v with respect to the basis $\mathcal{B}$ and we denote by

$$[v]_{\mathcal{B}} = \begin{pmatrix} v^1 \\ \vdots \\ v^n \end{pmatrix}$$

the **coordinate vector** of v with respect to $\mathcal{B}$.
(3) If we know that $\dim V = n$, then:
   (a) More than n vectors in V must be linearly dependent;
   (b) Fewer than n vectors in V cannot span V;
   (c) Any n linearly independent vectors span V;
   (d) Any n vectors that span V must be linearly independent;
   (e) If k vectors span V, then $k \geq n$ and some subset of those k vectors must be a basis of V;
   (f) If a set of m vectors is linearly independent, then $m \leq n$ and we can always complete that set to form a basis of V.

▶ From Chap. 3 onwards, our vector spaces are all finite-dimensional, hence here we concentrate on finite spanning sets and finite bases. Infinite bases and, correspondingly, infinite-dimensional vector spaces would bring us to the realm of Functional Analysis.

Example 2.26 The vectors b_1, b_2, b_3 in Example 2.17 form a basis of $\mathbb{R}^3$ since they are linearly independent and they are exactly as many as the dimension of $\mathbb{R}^3$. □

Example 2.27 (Gauss–Jordan Elimination) We are going to compute here the coordinates of $v = \begin{bmatrix} 1 \\ 1 \\ 1 \end{bmatrix}$ with respect to the basis $\mathcal{B} := \{b_1, b_2, b_3\}$ from

Example 2.17. The seeked coordinates $[v]_B = \begin{pmatrix} v^1 \\ \vdots \\ v^n \end{pmatrix}$ must satisfy the equation

$$v^1 \begin{bmatrix} 1 \\ 2 \\ 3 \end{bmatrix} + v^2 \begin{bmatrix} 4 \\ 5 \\ 6 \end{bmatrix} + v^3 \begin{bmatrix} 7 \\ 8 \\ 0 \end{bmatrix} = \begin{bmatrix} 1 \\ 1 \\ 1 \end{bmatrix},$$

so to find them we have to solve the following system of linear equations:

$$\begin{cases} v^1 + 4v^2 + 7v^3 = 1 \\ 2v^1 + 5v^2 + 8v^3 = 1 \\ 3v^1 + 6v^2 \quad\quad\;\; = 1. \end{cases}$$

For that purpose, we may equivalently reduce the following augmented matrix

$$\begin{bmatrix} 1 & 4 & 7 & | & 1 \\ 2 & 5 & 8 & | & 1 \\ 3 & 6 & 0 & | & 1 \end{bmatrix}$$

to echelon form, using the Gauss–Jordan elimination method. We are going to perform both calculations in parallel, which will also point out that they are indeed seemingly different incarnations of the same method.

By multiplying the first equation/row by 2 (resp. 3) and subtracting it from the second (resp. third) equation/row we obtain

$$\begin{cases} v^1 + 4v^2 + 7v^3 = 1 \\ \quad\;\; -3v^2 - 6v^3 = -1 \\ \quad\;\; -6v^2 - 21v^3 = -2 \end{cases} \quad\leftrightsquigarrow\quad \begin{bmatrix} 1 & 4 & 7 & | & 1 \\ 0 & -3 & -6 & | & -1 \\ 0 & -6 & -21 & | & -2 \end{bmatrix}.$$

By multiplying the second equation/row by $-\frac{1}{3}$ and by adding to the first (resp. third) equation/row the second equation/row multiplied by $\frac{4}{3}$ (resp. -2) we obtain

$$\begin{cases} v^1 \quad\quad\;\; - v^3 = -\frac{1}{3} \\ \quad\;\; v^2 + 2v^3 = \frac{1}{3} \\ \quad\quad\;\; -9v^3 = 0 \end{cases} \quad\leftrightsquigarrow\quad \begin{bmatrix} 1 & 0 & -1 & | & -\frac{1}{3} \\ 0 & 1 & 2 & | & \frac{1}{3} \\ 0 & 0 & -9 & | & 0 \end{bmatrix}.$$

2.2 Bases

The last equation/row shows that $v^3 = 0$, hence by backward substitution we obtain the solution

$$\begin{cases} v^1 & = -\frac{1}{3} \\ v^2 & = \frac{1}{3} \\ v^3 & = 0 \end{cases} \quad \longleftrightarrow \quad \left[\begin{array}{ccc|c} 1 & 0 & 0 & -\frac{1}{3} \\ 0 & 1 & 0 & \frac{1}{3} \\ 0 & 0 & 1 & 0 \end{array}\right].$$

□

Example 2.28 When $V = \mathbb{R}^n$ and $\mathcal{B} = \mathcal{E}$ is the standard basis, the coordinate vector $v \in \mathbb{R}^n$ coincides with the vector itself! In this very special case, we have $[v]_\mathcal{E} = v$. □

EXERCISE 2.29. Let

$$\mathcal{B} = \left\{ \begin{bmatrix} 1 \\ 0 \\ 0 \end{bmatrix}, \begin{bmatrix} 1 \\ 1 \\ 0 \end{bmatrix}, \begin{bmatrix} 1 \\ 1 \\ 1 \end{bmatrix} \right\}.$$

(1) Show that $\mathcal{B}$ is a basis of $\mathbb{R}^3$.
(2) Determine the coordinate vector, $[v]_\mathcal{B}$, of

$$v = \begin{bmatrix} 0 \\ 1 \\ \pi \end{bmatrix}$$

with respect to $\mathcal{B}$.
(3) Determine the vector $w \in \mathbb{R}^3$ that has coordinate vector

$$[w]_\mathcal{B} = \begin{pmatrix} 1 \\ 2 \\ 3 \end{pmatrix}.$$

EXERCISE 2.30. Let V be the vector space consisting of all 2×2 matrices with zero trace, namely

$$V := \left\{ \begin{bmatrix} a & b \\ c & d \end{bmatrix} : a, b, c, d \in \mathbb{R} \text{ and } a + d = 0 \right\}.$$

(1) Show that

$$\mathcal{B} := \left\{ \underbrace{\begin{bmatrix} 1 & 0 \\ 0 & -1 \end{bmatrix}}_{b_1}, \underbrace{\begin{bmatrix} 0 & 1 \\ 0 & 0 \end{bmatrix}}_{b_2}, \underbrace{\begin{bmatrix} 0 & 0 \\ 1 & 0 \end{bmatrix}}_{b_3} \right\}$$

is a basis of V.

(2) Show that

$$\tilde{\mathcal{B}} := \left\{ \underbrace{\begin{bmatrix} 1 & 0 \\ 0 & -1 \end{bmatrix}}_{\tilde{b}_1}, \underbrace{\begin{bmatrix} 0 & -1 \\ 1 & 0 \end{bmatrix}}_{\tilde{b}_2}, \underbrace{\begin{bmatrix} 0 & 1 \\ 1 & 0 \end{bmatrix}}_{\tilde{b}_3} \right\}$$

is another basis of V.

(3) Compute the coordinates of

$$v = \begin{bmatrix} 2 & 1 \\ 7 & -2 \end{bmatrix}$$

with respect to $\mathcal{B}$ and with respect to $\tilde{\mathcal{B}}$.

2.3 The Einstein Convention

2.3.1 A Convenient Summation Convention

We start by setting a notation that will turn out to be useful later on. Recall that if $\mathcal{B} = \{b_1, b_2, b_3\}$ is a basis of a vector space V, any vector $v \in V$ can be written as

$$v = v^1 b_1 + v^2 b_2 + v^3 b_3 \tag{2.1}$$

for appropriate $v^1, v^2, v^3 \in \mathbb{R}$.

Notation From now on, expressions like the one in Eq. 2.1 will be written as

$$v = \cancel{v^1 b_1 + v^2 b_2 + v^3 b_3} = v^j b_j . \tag{2.2}$$

That is, from now on when an index appears *twice – once as a subscript and once as a superscript –* in a term, we know that it means that there is a summation over all possible values of that index. The summation symbol will not be displayed.

2.3 The Einstein Convention

On the other hand, indices that are not repeated in expressions like $a_{ij}x^k y^j$ are *free indices* not subject to summation.

Examples 2.31 For indices ranging over $\{1, 2, 3\}$, i.e., $n = 3$:

(1) The expression $a_{ij}x^i y^k$ means

$$a_{ij}x^i y^k = a_{1j}x^1 y^k + a_{2j}x^2 y^k + a_{3j}x^3 y^k,$$

and could be called R_j^k (meaning that R_j^k and $a_{ij}x^i y^k$ both depend on the indices j and k).

(2) Likewise,

$$a_{ij}x^k y^j = a_{i1}x^k y^1 + a_{i2}x^k y^2 + a_{i3}x^k y^3 =: Q_i^k.$$

(3) Further

$$a_{ij}x^i y^j = a_{11}x^1 y^1 + a_{12}x^1 y^2 + a_{13}x^1 y^3$$
$$+ a_{21}x^2 y^1 + a_{22}x^2 y^2 + a_{23}x^2 y^3$$
$$+ a_{31}x^3 y^1 + a_{32}x^3 y^2 + a_{33}x^3 y^3 =: P$$

(4) An expression like

$$A^i B_{k\ell}^j C^\ell =: D_k^{ij}$$

makes sense. Here the indices i, j, k are free (i.e., free to range in $\{1, 2, \ldots, n\}$) and ℓ is a summation index.

(5) On the other hand an expression like

$$E_{ij} F_\ell^{jk} G^\ell = H_i^{jk}$$

does *not* make sense because the expression on the left has only two free indices, i and k, while j and ℓ are summation indices and neither of them can appear on the right hand side.

Notation Since $v^j b_j$ denotes a sum, we choose to denote the indices of the generic term of a sum with *capital letters*. For example, we write $v^I b_I$ and the above expressions could have been written as

(1)

$$a_{ij}x^i y^k = \sum_{I=1}^{3} a_{Ij}x^I y^K = a_{1j}x^1 y^k + a_{2j}x^2 y^k + a_{3j}x^3 y^k,$$

(2)
$$a_{ij}x^k y^j = \sum_{J=1}^{3} a_{iJ}x^K y^J = a_{i1}x^k y^1 + a_{i2}x^k y^2 + a_{i3}x^k y^3.$$

(3)
$$a_{ij}x^i y^j = \sum_{J=1}^{3}\sum_{I=1}^{3} a_{IJ}x^I y^J =$$
$$= a_{11}x^1 y^1 + a_{12}x^1 y^2 + a_{13}x^1 y^3$$
$$+ a_{21}x^2 y^1 + a_{22}x^2 y^2 + a_{23}x^2 y^3$$
$$+ a_{31}x^3 y^1 + a_{32}x^3 y^2 + a_{33}x^3 y^3.$$

$\square$

EXERCISE 2.32. Let
$$A = (A^i_j) \in \mathbb{R}^{\ell \times m} \quad \text{and} \quad B = (B^i_j) \in \mathbb{R}^{m \times n}$$
be matrices, where the upper indices are the row indices and the lower indices are the column indices. Moreover, let
$$x = (x^i) \in \mathbb{R}^\ell \quad \text{and} \quad y = (y^i) \in \mathbb{R}^n$$
be column vectors. How are the coordinates of the following expressions written using the Einstein convention?

(1) AB
(2) By
(3) $y^T B^T$
(4) $xy^T B^T$

2.3.2 Change of Basis

Let $\mathcal{B}$ and $\widetilde{\mathcal{B}}$ be two bases of a vector space V and let

$$L := L_{\mathcal{B}\widetilde{\mathcal{B}}} = \begin{bmatrix} L^1_1 & \cdots & L^1_n \\ \vdots & & \vdots \\ L^n_1 & \cdots & L^n_n \end{bmatrix} \qquad (2.3)$$

2.3 The Einstein Convention

be the **matrix of the change of basis** from the "old" basis $\mathcal{B}$ to a "new" basis $\widetilde{\mathcal{B}}$. Recall that the entries of the j-th column of L are the coordinates of the new basis vector $\tilde{b}_j$ with respect to the old basis $\mathcal{B}$.

> **Mnemonic:** Upper indices go **up** to down, i.e., they are row indices.
> Lower indices go **left** to right, i.e., they are column indices.

With the Einstein convention we can write

$$\boxed{\tilde{b}_j = L^i_j b_i}, \tag{2.4}$$

or, equivalently,

$$\boxed{(\tilde{b}_1 \ \ldots \ \tilde{b}_n) = (b_1 \ \ldots \ b_n) L},$$

where we use some convenient formal notation: The multiplication is to be performed with the usual rules for vectors and matrices, though, in this case, the entries of the row vectors $(\tilde{b}_1 \ \ldots \ \tilde{b}_n)$ and $(b_1 \ \ldots \ b_n)$ are not real numbers but vectors themselves.

EXERCISE 2.33. Let $\mathcal{A}$, $\mathcal{B}$ and $\mathcal{C}$ be three bases of a vector space V, let $L_{\mathcal{AB}}$ be the matrix of the change of basis from $\mathcal{A}$ to $\mathcal{B}$, let $L_{\mathcal{BC}}$ be the matrix of the change of basis from $\mathcal{B}$ to $\mathcal{C}$, and let $L_{\mathcal{AC}}$ be the matrix of the change of basis from $\mathcal{A}$ to $\mathcal{C}$. Show that then we have

$$L_{\mathcal{AC}} = L_{\mathcal{AB}} L_{\mathcal{BC}}.$$

If $\Lambda = L^{-1}$ denotes the matrix of the change of basis from $\widetilde{\mathcal{B}}$ to $\mathcal{B}$, then, using the same formal notation as above, we have

$$\boxed{(b_1 \ \ldots \ b_n) = (\tilde{b}_1 \ \ldots \ \tilde{b}_n) \Lambda}.$$

Equivalently, this can be written in compact form using the Einstein notation as

$$\boxed{b_j = \Lambda^i_j \tilde{b}_i}.$$

Remark 2.34 Mathematically, the set of all matrices of change of basis for a vector space V equipped with matrix multiplication forms a *group*, called the *general linear group* of V. Key here is the relation $L_{\mathcal{AC}} = L_{\mathcal{AB}} L_{\mathcal{BC}}$ from Exercise 2.33 and the relation $L_{\mathcal{BA}} = L_{\mathcal{AB}}^{-1}$.

The corresponding relations for the vector coordinates are

$$\begin{pmatrix} v^1 \\ \vdots \\ v^i \\ \vdots \\ v^n \end{pmatrix} = \begin{bmatrix} L_1^1 & \cdots\cdots & L_n^1 \\ \vdots & & \vdots \\ L_1^i & \cdots\cdots & L_n^i \\ \vdots & & \vdots \\ L_1^n & \cdots\cdots & L_n^n \end{bmatrix} \begin{pmatrix} \tilde{v}^1 \\ \vdots \\ \tilde{v}^i \\ \vdots \\ \tilde{v}^n \end{pmatrix} \quad \text{and} \quad \begin{pmatrix} \tilde{v}^1 \\ \vdots \\ \tilde{v}^i \\ \vdots \\ \tilde{v}^n \end{pmatrix} = \begin{bmatrix} \Lambda_1^1 & \cdots\cdots & \Lambda_n^1 \\ \vdots & & \vdots \\ \Lambda_1^i & \cdots\cdots & \Lambda_n^i \\ \vdots & & \vdots \\ \Lambda_1^n & \cdots\cdots & \Lambda_n^n \end{bmatrix} \begin{pmatrix} v^1 \\ \vdots \\ v^i \\ \vdots \\ v^n \end{pmatrix}$$

and these can be written with the Einstein convention respectively as

$$\boxed{v^i = L^i_j \tilde{v}^j} \quad \text{and} \quad \boxed{\tilde{v}^i = \Lambda^i_j v^j}, \tag{2.5}$$

or, in matrix notation,

$$[v]_\mathcal{B} = L_{\mathcal{B}\tilde{\mathcal{B}}} [v]_{\tilde{\mathcal{B}}} \quad \text{and} \quad [v]_{\tilde{\mathcal{B}}} = (L_{\mathcal{B}\tilde{\mathcal{B}}})^{-1} [v]_\mathcal{B} = L_{\tilde{\mathcal{B}}\mathcal{B}} [v]_\mathcal{B}.$$

▶ The coordinate vectors change in a way *opposite* to the basis change. Hence, we say that the coordinate vectors are **contravariant**, because they change by L^{-1} when the basis changes by L.

Example 2.35 We consider the following two bases of $\mathbb{R}^2$

$$\mathcal{B} = \left\{ \underbrace{\begin{bmatrix} 1 \\ 0 \end{bmatrix}}_{b_1}, \underbrace{\begin{bmatrix} 2 \\ 1 \end{bmatrix}}_{b_2} \right\}$$

$$\tilde{\mathcal{B}} = \left\{ \underbrace{\begin{bmatrix} 3 \\ 1 \end{bmatrix}}_{\tilde{b}_1}, \underbrace{\begin{bmatrix} -1 \\ -1 \end{bmatrix}}_{\tilde{b}_2} \right\}$$

(2.6)

and we look for the matrix of the change of basis. Namely, we look for a matrix L such that

$$(\tilde{b}_1 \; \tilde{b}_2) = (b_1 \; b_2) \, L,$$

that is

$$\begin{bmatrix} 3 & -1 \\ 1 & -1 \end{bmatrix} = \begin{bmatrix} 1 & 2 \\ 0 & 1 \end{bmatrix} L.$$

2.3 The Einstein Convention

There are two alternative ways of finding L:

(1) *With matrix inversion:* Recall that

$$\begin{bmatrix} a & b \\ c & d \end{bmatrix}^{-1} = \frac{1}{D} \begin{bmatrix} d & -b \\ -c & a \end{bmatrix}, \qquad (2.7)$$

where $D = \det \begin{bmatrix} a & b \\ c & d \end{bmatrix}$ is the *determinant* (see also Sect. 2.4.2). Thus

$$L = \begin{bmatrix} 1 & 2 \\ 0 & 1 \end{bmatrix}^{-1} \begin{bmatrix} 3 & -1 \\ 1 & -1 \end{bmatrix} = \begin{bmatrix} 1 & -2 \\ 0 & 1 \end{bmatrix} \begin{bmatrix} 3 & -1 \\ 1 & -1 \end{bmatrix} = \begin{bmatrix} 1 & 1 \\ 1 & -1 \end{bmatrix}.$$

(2) *With Gauss-Jordan elimination:*

$$\begin{bmatrix} 1 & 2 & 3 & -1 \\ 0 & 1 & 1 & -1 \end{bmatrix} \rightsquigarrow \begin{bmatrix} 1 & 0 & 1 & 1 \\ 0 & 1 & 1 & -1 \end{bmatrix}$$

□

2.3.3 The Kronecker Delta Symbol

Notation The **Kronecker delta symbol** δ^i_j is defined as

$$\delta^i_j := \begin{cases} 1 & \text{if } i = j \\ 0 & \text{if } i \neq j. \end{cases} \qquad (2.8)$$

Similarly, we define δ_{ij} and δ^{ij}.

Examples 2.36 If L is a matrix, the (i, j)-entry of L is the coefficient in the i-th row and j-th column, and is denoted by L^i_j.

(1) The $n \times n$ **identity matrix**

$$I = \begin{bmatrix} 1 & \cdots & 0 \\ \vdots & \ddots & \vdots \\ 0 & \cdots & 1 \end{bmatrix}$$

has (i, j)-entry equal to δ^i_j.

(2) Let L and M be two square matrices. The (i,j)-th entry of the product

$$ML = \begin{bmatrix} M_1^1 & \cdots\cdots & M_n^1 \\ \vdots & & \vdots \\ M_1^i & \cdots\cdots & M_n^i \\ \vdots & & \vdots \\ M_1^n & \cdots\cdots & M_n^n \end{bmatrix} \begin{bmatrix} L_1^1 . L_j^1 . L_n^1 \\ \vdots & \vdots & \vdots \\ \vdots & \vdots & \vdots \\ L_1^n . L_j^n . L_n^n \end{bmatrix}$$

equals the *dot product*[1] of the i-th row of M and j-th column of L,

$$\begin{pmatrix} M_1^i & \cdots & M_n^i \end{pmatrix} \cdot \begin{pmatrix} L_j^1 \\ \vdots \\ L_j^n \end{pmatrix} = M_1^i L_j^1 + \cdots + M_n^i L_j^n,$$

or, using the Einstein convention,

$$M_k^i L_j^k.$$

Notice that, since in general $ML \neq LM$, it follows that

$$M_k^i L_j^k \neq L_k^i M_j^k = M_j^k L_k^i.$$

However, in the special case where $M = \Lambda = L^{-1}$, we have $\Lambda L = L\Lambda = I$ and here we can write

$$\Lambda_k^i L_j^k = \delta_j^i = L_k^i \Lambda_j^k.$$

$\square$

Remark 2.37 Using the Kronecker delta symbol, we can check that the notations in Eqs. 2.5 are all consistent. In fact, we should have

$$v^i b_i = v = \tilde{v}^i \tilde{b}_i, \tag{2.9}$$

and indeed, by Eqs. 2.5, we have

$$\tilde{v}^i \tilde{b}_i = \left(\Lambda_j^i v^j\right)\left(L_i^k b_k\right) = \left(\Lambda_j^i L_i^k\right) v^j b_k = \delta_j^k v^j b_k = v^j b_j,$$

where the equality $\Lambda_j^i L_i^k = \delta_j^k$ amounts to $\Lambda = L^{-1}$.

---

[1] The definition of dot product is recalled in Example 3.17.

2.4 Linear Transformations

Two words of warning:

- The two expressions $v^j b_j$ and $v^k b_k$ are identical, as the indices j and k are *dummy indices*, that is, can be replaced by other symbols throughout, without changing the meaning of the expression (as long as the symbols do not collide with other symbols already used in the expression).
- When multiplying two different expressions in Einstein notation, we should be careful to distinguish by different letters different summation indices. For example, if $\tilde{v}^i = \Lambda^i_j v^j$ and $\tilde{b}_i = L^j_i b_j$, in order to perform the multiplication $\tilde{v}^i \tilde{b}_i$ we have to make sure to replace one of the dummy indices in the two expressions. So, for example, we can write $\tilde{b}_i = L^k_i b_k$, so that $\tilde{v}^i \tilde{b}_i = \Lambda^i_j v^j L^k_i b_k$.

EXERCISE 2.38. This is a continuation of Exercise 2.32. Use the Kronecker delta symbol to write the coordinates of $A^T x$ according to the Einstein convention.

2.4 Linear Transformations

2.4.1 Linear Transformations as (1, 1)-Tensors

Recall that a linear transformation from V to itself, $T : V \to V$, is a function (or *map* or *transformation*) that satisfies the property

$$T(\alpha v + \beta w) = \alpha T(v) + \beta T(w),$$

for all $\alpha, \beta \in \mathbb{R}$ and all $v, w \in V$. Once we choose a basis $\mathcal{B} = \{b_1, \ldots b_n\}$ of V, the transformation T is represented by a matrix A, called the **matrix of the linear transformation with respect to that basis**. The columns of A are the coordinate vectors of $T(b_1), \ldots, T(b_n)$ with respect to $\mathcal{B}$. Then that matrix A gives the effect of T on coordinate vectors as follows: If $T(v)$ is the value of the transformation T on the vector v, with respect to a basis $\mathcal{B}$ we have that

$$[v]_\mathcal{B} \longmapsto [T(v)]_\mathcal{B} = A[v]_\mathcal{B} . \tag{2.10}$$

If $\widetilde{\mathcal{B}}$ is another basis, we have also

$$[v]_{\widetilde{\mathcal{B}}} \longmapsto [T(v)]_{\widetilde{\mathcal{B}}} = \widetilde{A}[v]_{\widetilde{\mathcal{B}}}, \tag{2.11}$$

where now $\widetilde{A}$ is the matrix of the transformation T with respect to the basis $\widetilde{\mathcal{B}}$.

We want to find now the relation between A and $\widetilde{A}$. Let $L := L_{\mathcal{B}\widetilde{\mathcal{B}}}$ be the matrix of the change of basis from $\mathcal{B}$ to $\widetilde{\mathcal{B}}$. Then, for any $v \in V$,

$$[v]_{\widetilde{\mathcal{B}}} = L^{-1}[v]_\mathcal{B} . \tag{2.12}$$

In particular the above equation holds for the vector $T(v)$, that is

$$[T(v)]_{\widetilde{\mathcal{B}}} = L^{-1}[T(v)]_{\mathcal{B}}. \tag{2.13}$$

Then we have

$$\widetilde{A}L^{-1}[v]_{\mathcal{B}} \stackrel{(2.12)}{=} \widetilde{A}[v]_{\widetilde{\mathcal{B}}} \stackrel{(2.11)}{=} [T(v)]_{\widetilde{\mathcal{B}}} \stackrel{(2.13)}{=} L^{-1}[T(v)]_{\mathcal{B}} \stackrel{(2.10)}{=} L^{-1}A[v]_{\mathcal{B}}$$

for every vector $v \in V$. If follows that $\widetilde{A}L^{-1} = L^{-1}A$ or, equivalently,

$$\widetilde{A} = L^{-1}AL, \tag{2.14}$$

which, in Einstein notation, reads

$$\widetilde{A}^i_j = \Lambda^i_k A^k_m L^m_j.$$

*We say that the linear transformation T is a **tensor of type** $(1, 1)$.*[2]

Example 2.39 Let $V = \mathbb{R}^2$ and let $\mathcal{B}$ and $\widetilde{\mathcal{B}}$ be the bases in Example 2.35. The matrices corresponding to the change of coordinates are

$$L := L_{\mathcal{B}\widetilde{\mathcal{B}}} = \begin{bmatrix} 1 & 1 \\ 1 & -1 \end{bmatrix} \quad \text{and} \quad L^{-1} = \frac{1}{-2}\begin{bmatrix} -1 & -1 \\ -1 & 1 \end{bmatrix} = \begin{bmatrix} \frac{1}{2} & \frac{1}{2} \\ \frac{1}{2} & -\frac{1}{2} \end{bmatrix},$$

where in the last equality we used the formula for the inverse of a matrix in (2.7).

Let $T : \mathbb{R}^2 \to \mathbb{R}^2$ be the linear transformation that in the basis $\mathcal{B}$ takes the form

$$A = \begin{bmatrix} 1 & 3 \\ 2 & 4 \end{bmatrix}.$$

Then according to Eq. 2.14 the matrix $\widetilde{A}$ of the linear transformation T with respect to the basis $\widetilde{\mathcal{B}}$ is

$$\widetilde{A} = L^{-1}AL = \begin{bmatrix} \frac{1}{2} & \frac{1}{2} \\ \frac{1}{2} & -\frac{1}{2} \end{bmatrix}\begin{bmatrix} 1 & 3 \\ 2 & 4 \end{bmatrix}\begin{bmatrix} 1 & 1 \\ 1 & -1 \end{bmatrix} = \begin{bmatrix} 5 & -2 \\ -1 & 0 \end{bmatrix}.$$

□

---

[2] See Sect. 5.2 for an explanation of the terminology.

2.4 Linear Transformations

Example 2.40 We now look for the **standard matrix** of a linear transformation T, that is, the matrix M that represents T with respect to the standard basis of $\mathbb{R}^2$, which we denote by

$$\mathcal{E} := \left\{ \underbrace{\begin{bmatrix} 1 \\ 0 \end{bmatrix}}_{e_1}, \underbrace{\begin{bmatrix} 0 \\ 1 \end{bmatrix}}_{e_2} \right\}.$$

We want to apply again Eq. 2.14 and, hence, we first need to find the matrix $S := L_{\mathcal{E}\mathcal{B}}$ of the change of basis from $\mathcal{E}$ to $\mathcal{B}$. Recall that the columns of S are the coordinates of b_j with respect to the basis $\mathcal{E}$, that is

$$S = \begin{bmatrix} 1 & 2 \\ 0 & 1 \end{bmatrix}.$$

According to Eq. 2.14, we have

$$A = S^{-1}MS,$$

from which, using again Eq. 2.7, we obtain

$$M = SAS^{-1} = \begin{bmatrix} 1 & 2 \\ 0 & 1 \end{bmatrix} \begin{bmatrix} 1 & 3 \\ 2 & 4 \end{bmatrix} \begin{bmatrix} 1 & -2 \\ 0 & 1 \end{bmatrix} = \begin{bmatrix} 1 & 2 \\ 0 & 1 \end{bmatrix} \begin{bmatrix} 1 & 1 \\ 2 & 0 \end{bmatrix} = \begin{bmatrix} 5 & 1 \\ 2 & 0 \end{bmatrix}.$$

□

Example 2.41 Let $V := P_2(\mathbb{R})$ be the vector space of polynomials of degree ≤ 2, and let $T : P_2(\mathbb{R}) \to P_2(\mathbb{R})$ be the linear transformation given by differentiating a polynomial and then multiplying the derivative by x,

$$T(p(x)) := xp'(x),$$

so that $T(a + bx + cx^2) = x(b + 2cx) = bx + 2cx^2$. Let

$$\mathcal{B} := \{1, x, x^2\} \quad \text{and} \quad \widetilde{\mathcal{B}} := \{x, x - 1, x^2 - 1\}$$

be two bases of $P_2(\mathbb{R})$. Since

$$T(1) = 0 = 0 \cdot 1 + 0 \cdot x + 0 \cdot x^2$$
$$T(x) = x = 0 \cdot 1 + 1 \cdot x + 0 \cdot x^2$$
$$T(x^2) = 2x^2 = 0 \cdot 1 + 0 \cdot x + 2 \cdot x^2$$

and

$$T(x) = x = 1 \cdot x + 0 \cdot (x-1) + 0 \cdot (x^2 - 1)$$
$$T(x-1) = x = 1 \cdot x + 0 \cdot (x-1) + 0 \cdot (x^2 - 1)$$
$$T(x^2 - 1) = 2x^2 = 2 \cdot x - 2 \cdot (x-1) + 2 \cdot (x^2 - 1),$$

then

$$A = \begin{bmatrix} 0 & 0 & 0 \\ 0 & 1 & 0 \\ 0 & 0 & 2 \end{bmatrix} \quad \text{and} \quad \widetilde{A} = \begin{bmatrix} 1 & 1 & 2 \\ 0 & 0 & -2 \\ 0 & 0 & 2 \end{bmatrix}.$$

One can check that indeed $AL = L\widetilde{A}$ or, equivalently $\widetilde{A} = L^{-1}AL$, where

$$L = \begin{bmatrix} 0 & -1 & -1 \\ 1 & 1 & 0 \\ 0 & 0 & 1 \end{bmatrix}$$

is the matrix of the change of basis. □

▶ Geometric features often point out a basis particularly well-suited to tackle a given problem. That observation, combined with the transformation law in Eq. 2.14, yields a good strategy for finding the matrix of a linear transformation with respect to a certain basis, as the following example illustrates. This viewpoint will be systematized in Sect. 2.4.3.

Example 2.42 Let $T : \mathbb{R}^3 \to \mathbb{R}^3$ be the *orthogonal projection* onto the plane $\mathcal{P}$ of equation

$$2x + y - z = 0.$$

This means that the transformation T is characterized by the fact that

- it does not change vectors in the plane $\mathcal{P}$, and
- it sends vectors perpendicular to $\mathcal{P}$ to the zero vector in $\mathcal{P}$.

We want to find the standard matrix for T. As suggested above, we compute the matrix of T with respect to a basis $\mathcal{B}$ of $\mathbb{R}^3$ well adapted to the problem, then use Eq. 2.14 after having found the matrix $L_{\mathcal{E}\mathcal{B}}$ of the change of basis.

2.4 Linear Transformations

To this purpose, we choose two linearly independent vectors in the plane $\mathcal{P}$ and a third vector perpendicular to $\mathcal{P}$. For instance, we set

$$\mathcal{B} := \left\{ \underbrace{\begin{bmatrix} 1 \\ 0 \\ 2 \end{bmatrix}}_{b_1}, \underbrace{\begin{bmatrix} 0 \\ 1 \\ 1 \end{bmatrix}}_{b_2}, \underbrace{\begin{bmatrix} 2 \\ 1 \\ -1 \end{bmatrix}}_{b_3} \right\},$$

where the coordinates of b_1 and b_2 satisfy the equation of the plane, while the coordinates of b_3 are the coefficients of the equation describing $\mathcal{P}$. Let $\mathcal{E}$ be the standard basis of $\mathbb{R}^3$.

Since

$$T(b_1) = b_1, \qquad T(b_2) = b_2 \qquad \text{and } T(b_3) = 0,$$

the matrix of T with respect to $\mathcal{B}$ is simply

$$A = \begin{bmatrix} 1 & 0 & 0 \\ 0 & 1 & 0 \\ 0 & 0 & 0 \end{bmatrix}, \qquad (2.15)$$

where we recall that the j-th column is the coordinate vector $[T(b_j)]_\mathcal{B}$ of the vector $T(b_j)$ with respect to the basis $\mathcal{B}$.

The matrix of the change of basis from $\mathcal{E}$ to $\mathcal{B}$ is

$$L = \begin{bmatrix} 1 & 0 & 2 \\ 0 & 1 & 1 \\ 2 & 1 & -1 \end{bmatrix},$$

hence, by Gauss–Jordan elimination,

$$L^{-1} = \begin{bmatrix} \frac{1}{3} & -\frac{1}{3} & \frac{1}{3} \\ -\frac{1}{3} & \frac{5}{6} & \frac{1}{6} \\ \frac{1}{3} & \frac{1}{6} & -\frac{1}{6} \end{bmatrix}.$$

Therefore

$$M = LAL^{-1} = \cdots = \begin{bmatrix} \frac{1}{3} & -\frac{1}{3} & \frac{1}{3} \\ -\frac{1}{3} & \frac{5}{6} & \frac{1}{6} \\ \frac{1}{3} & \frac{1}{6} & \frac{5}{6} \end{bmatrix}.$$

□

2.4.2 Conjugate Matrices

The above calculations can be summarized by the *commutativity* of the following diagram. Here, the vertical arrows correspond to the operation of change of basis from $\mathcal{B}$ to $\widetilde{\mathcal{B}}$ (recall that the coordinate vectors are contravariant tensors, that is, they transform as $[v]_{\widetilde{\mathcal{B}}} = L^{-1}[v]_{\mathcal{B}}$) and the horizontal arrows correspond to the operation of applying the transformation T with respect to the two different basis:

$$\begin{array}{ccc} [v]_{\mathcal{B}} & \xrightarrow{A} & [T(v)]_{\mathcal{B}} \\ {\scriptstyle L^{-1}}\downarrow & & \downarrow{\scriptstyle L^{-1}} \\ [v]_{\widetilde{\mathcal{B}}} & \xrightarrow{\widetilde{A}} & [T(v)]_{\widetilde{\mathcal{B}}} \end{array}$$

Saying that **the diagram is commutative** is saying that if one starts from the upper left hand corner, reaching the lower right hand corner following either one of the two paths (i.e., either first to the right via A then down via L^{-1}, or first down via L^{-1} and then right via $\widetilde{A}$) has exactly the same effect. In other words, changing coordinates first then applying the transformation T yields exactly the same affect as applying first the transformation T and then the change of coordinates, that is, $L^{-1}A = \widetilde{A}L^{-1}$ or, equivalently,

$$\widetilde{A} = L^{-1}AL.$$

In this case we say that A and $\widetilde{A}$ are *conjugate* matrices. This means that A and $\widetilde{A}$ represent the same transformation with respect to different bases.

> DEFINITION 2.43. We say that two matrices A and $\widetilde{A}$ are **conjugate** if there exists an invertible matrix L such that $\widetilde{A} = L^{-1}AL$.

Example 2.44 The three matrices from Examples 2.39 and 2.40

$$A = \begin{bmatrix} 1 & 3 \\ 2 & 4 \end{bmatrix} \quad M = \begin{bmatrix} 5 & 1 \\ 2 & 0 \end{bmatrix} \quad \text{and} \quad \widetilde{A} = \begin{bmatrix} 5 & -2 \\ -1 & 0 \end{bmatrix}$$

are all conjugate. Indeed, we have

$$\widetilde{A} = L^{-1}AL, \qquad A = S^{-1}MS \quad \text{and} \quad \widetilde{A} = R^{-1}MR,$$

where L and S may be found in those examples and where $R := SL$. □

We now review some facts about conjugate matrices. Recall that the **characteristic polynomial** of a square matrix A is the polynomial

$$p_A(\lambda) := \det(A - \lambda I).$$

2.4 Linear Transformations

Let us assume that A and $\widetilde{A}$ are conjugate matrices, that is $\widetilde{A} = L^{-1}AL$ for some invertible matrix L. Then

$$\begin{aligned} p_{\widetilde{A}}(\lambda) &= \det(\widetilde{A} - \lambda I) = \det(L^{-1}AL - \lambda L^{-1}IL) \\ &= \det(L^{-1}(A - \lambda I)L) \\ &= (\det L^{-1})\det(A - \lambda I)(\det L) = p_A(\lambda), \end{aligned} \qquad (2.16)$$

which means that any two conjugate matrices have the same characteristic polynomial.

Recall that the **eigenvalues** of a matrix A are the roots of its characteristic polynomial and we here usually allow complex roots. Then, by the *Fundamental Theorem of Algebra*, each $n \times n$ matrix has n (real or complex) eigenvalues counted with multiplicities as polynomial roots. Recall also the definitions of *determinant* and *trace* of a square matrix. By analysing the characteristic polynomial, we see that

(1) the **determinant** of a matrix is equal to the product of its eigenvalues (multiplied with multiplicities), and
(2) the **trace** of a matrix is equal to the sum of its eigenvalues (added with multiplicities).

From the computation in (2.16), it follows that, if the matrices A and $\widetilde{A}$ are conjugate, then:

- A and $\widetilde{A}$ have the same size;
- the eigenvalues of A (as well as their multiplicities) are the same as those of $\widetilde{A}$;
- $\det A = \det \widetilde{A}$;
- $\operatorname{tr} A = \operatorname{tr} \widetilde{A}$;
- A is invertible if and only if $\widetilde{A}$ is invertible.

Example 2.45 The matrices $A = \begin{bmatrix} 1 & 3 \\ 2 & 4 \end{bmatrix}$ and $A' = \begin{bmatrix} 1 & 2 \\ 2 & 4 \end{bmatrix}$ are not conjugate. In fact, A is invertible, as $\det A = -2 \neq 0$, while $\det A' = 0$, so that A' is not invertible. □

Example 2.46 In Example 2.42, we showed that the matrices

$$A = \begin{bmatrix} 1 & 0 & 0 \\ 0 & 1 & 0 \\ 0 & 0 & 0 \end{bmatrix} \quad \text{and} \quad M = \begin{bmatrix} \frac{1}{3} & -\frac{1}{3} & \frac{1}{3} \\ -\frac{1}{3} & \frac{5}{6} & \frac{1}{6} \\ \frac{1}{3} & \frac{1}{6} & \frac{5}{6} \end{bmatrix}$$

are conjugate. From this fact, we see most easily that $\det M = 0$ and $\operatorname{tr} M = 2$. □

2.4.3 Eigenbases

The possibility of choosing different bases is very important and often simplifies the calculations. Example 2.42 is such an example, where we choose an appropriate basis according to the specific problem. Other times, a basis can be chosen according to the symmetries and, completely at the opposite side, sometime there is just not a basis that is a preferred one. In the context of a linear transformation $T : V \to V$, a basis that is particularly convenient, when it exists, is an *eigenbasis* for that linear transformation.

Recall that an **eigenvector** of a linear transformation $T : V \to V$ is a vector $v \neq 0$ such that $T(v)$ is a multiple of v, say $T(v) = \lambda v$ and, in that case, the scaling number λ is called an **eigenvalue** of T. An **eigenbasis** is a basis of V consisting of eigenvectors of a linear transformation $T : V \to V$.

▶ The point of having an eigenbasis is that, with respect to this eigenbasis, the linear transformation is representable by a *diagonal* matrix, D. Hence, any other matrix representative of that linear transformation will be actually conjugate to a diagonal matrix D. Recall that diagonal matrices are extremely friendly for computations, so the possibility of producing eigenbases accounts for a main application of eigenvectors.

A linear transformation $T : V \to V$ for which an eigenbasis exists is then called **diagonalizable**.[3]

Given a linear transformation $T : V \to V$, in order to find an eigenbasis (assuming it exists) of T, we first represent T by some matrix A with respect to some chosen basis of V, and then perform the following steps:

(1) We find the eigenvalues by determining the roots of the characteristic polynomial of A (often allowing complex roots).
(2) For each eigenvalue λ, we find the corresponding eigenvectors by looking for the non-zero vectors in its **eigenspace**

$$E_\lambda := \ker(A - \lambda I).$$

When considering complex eigenvalues, the eigenspaces are determined as subspaces of the *complex vector space* $\mathbb{C}^n$. However, in this text, we concentrate on real cases.
(3) We determine whether there exists an eigenbasis.

---

[3] In general, when an eigenbasis does not exist, it is still possible to find a basis, with respect to which the linear transformation is as simple as possible, i.e., as close as possible to being diagonal. A **Jordan canonical form** provides such a best matrix representative of $T : V \to V$ and is necessarily conjugate to the first matrix representative A. In this text, we will not address these more general canonical forms.

2.4 Linear Transformations

We will illustrate this in the following examples.

Example 2.47 Let $T : \mathbb{R}^2 \to \mathbb{R}^2$ be the linear transformation given by the matrix $A = \begin{bmatrix} 3 & -4 \\ -4 & -3 \end{bmatrix}$ with respect to the standard basis of $\mathbb{R}^2$.

(1) The eigenvalues are the roots of the characteristic polynomial $p_\lambda(A)$. Since

$$p_A(\lambda) = \det(A - \lambda I) = \det \begin{bmatrix} 3 - \lambda & -4 \\ -4 & -3 - \lambda \end{bmatrix}$$

$$= (3 - \lambda)(-3 - \lambda) - 16 = \lambda^2 - 25 = (\lambda - 5)(\lambda + 5),$$

hence $\lambda = \pm 5$ are the eigenvalues of A.

(2) If λ is an eigenvalue of A, the eigenspace corresponding to λ is given by $E_\lambda = \ker(A - \lambda I)$. Note that

$$v \in E_\lambda \iff Av = \lambda v.$$

With our choice of A and with the resulting eigenvalues, we have

$$E_5 = \ker(A - 5I) = \ker \begin{bmatrix} -2 & -4 \\ -4 & -8 \end{bmatrix} = \text{span} \begin{bmatrix} 2 \\ -1 \end{bmatrix}$$

$$E_{-5} = \ker(A + 5I) = \ker \begin{bmatrix} 8 & -4 \\ -4 & 2 \end{bmatrix} = \text{span} \begin{bmatrix} 1 \\ 2 \end{bmatrix}.$$

1. The following is an eigenbasis for this linear transformation:

$$\tilde{\mathcal{B}} = \left\{ \tilde{b}_1 = \begin{bmatrix} 2 \\ -1 \end{bmatrix}, \tilde{b}_2 = \begin{bmatrix} 1 \\ 2 \end{bmatrix} \right\}$$

and

$$T(\tilde{b}_1) = 5\tilde{b}_1 = 5 \cdot \tilde{b}_1 + 0 \cdot \tilde{b}_2$$

$$T(\tilde{b}_2) = -5\tilde{b}_2 = 0 \cdot \tilde{b}_1 - 5 \cdot \tilde{b}_2,$$

so that $\tilde{A} = \begin{bmatrix} 5 & 0 \\ 0 & -5 \end{bmatrix}$ (Fig. 2.1).

Notice that the eigenspace E_5 consists of vectors on the line $x + 2y = 0$ and these vectors get scaled by the transformation T by a factor of 5. On the other hand, the eigenspace E_{-5} consists of vectors perpendicular to the line $x + 2y = 0$ and these vectors get flipped by the transformation T and then also scaled by a factor of 5. Hence T is just the reflection across the line $x + 2y = 0$ followed by multiplication by 5.

Fig. 2.1 Eigenbasis $\tilde{b}_1, \tilde{b}_2$ for the transformation T in Example 2.47

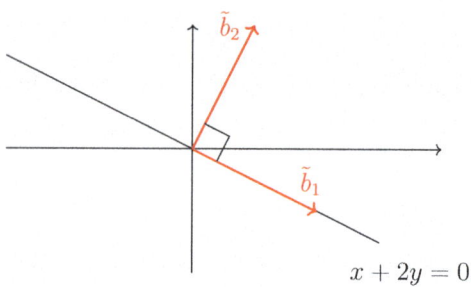

□

Example 2.48 Now let $T : \mathbb{R}^2 \to \mathbb{R}^2$ be the linear transformation given by the matrix $A = \begin{bmatrix} 1 & 2 \\ 4 & 3 \end{bmatrix}$ with respect to the standard basis of $\mathbb{R}^2$.

(1) The eigenvalues are the roots of the characteristic polynomial:

$$p_A(\lambda) = \det(A - \lambda I) = \det \begin{bmatrix} 1-\lambda & 2 \\ 4 & 3-\lambda \end{bmatrix}$$
$$= (1-\lambda)(3-\lambda) - 2 \cdot 4 = \lambda^2 - 4\lambda - 5 = (\lambda+1)(\lambda-5),$$

hence $\lambda = -1$ and $\lambda = 5$ are the eigenvalues of A.

(2) If λ is an eigenvalue of A, the eigenspace corresponding to λ is given by $E_\lambda = \ker(A - \lambda I)$. In this case we have

$$E_{-1} = \ker(A + I) = \ker \begin{bmatrix} 2 & 2 \\ 4 & 4 \end{bmatrix} = \text{span} \begin{bmatrix} 1 \\ -1 \end{bmatrix}$$

$$E_5 = \ker(A - 5I) = \ker \begin{bmatrix} -4 & 2 \\ 4 & -2 \end{bmatrix} = \text{span} \begin{bmatrix} 1 \\ 2 \end{bmatrix}.$$

(3) The following is an eigenbasis for this linear transformation:

$$\tilde{\mathcal{B}} = \left\{ \tilde{b}_1 = \begin{bmatrix} 1 \\ -1 \end{bmatrix}, \tilde{b}_2 = \begin{bmatrix} 1 \\ 2 \end{bmatrix} \right\}.$$

We have $T(\tilde{b}_1) = -\tilde{b}_1$ and $T(\tilde{b}_2) = 5\tilde{b}_2$, hence $\tilde{A} = \begin{bmatrix} -1 & 0 \\ 0 & 5 \end{bmatrix}$.

□

2.4 Linear Transformations

Example 2.49 Now let $T : \mathbb{R}^2 \to \mathbb{R}^2$ be the linear transformation given by the matrix $A = \begin{bmatrix} 5 & -3 \\ 3 & -1 \end{bmatrix}$ with respect to the standard basis of $\mathbb{R}^2$.

(1) The eigenvalues are the roots of the characteristic polynomial:

$$p_A(\lambda) = \det(A - \lambda I) = \det \begin{bmatrix} 5-\lambda & -3 \\ 3 & -1-\lambda \end{bmatrix}$$
$$= (5-\lambda)(-1-\lambda) + 9 = \lambda^2 - 4\lambda + 4 = (\lambda - 2)^2,$$

hence $\lambda = 2$ is the only eigenvalue of A.

(2) The eigenspace corresponding to $\lambda = 2$ is

$$E_2 = \ker(A - 2I) = \ker \begin{bmatrix} 3 & -3 \\ 3 & -3 \end{bmatrix} = \operatorname{span}\begin{bmatrix} 1 \\ 1 \end{bmatrix}.$$

(3) Since we cannot find two linearly independent eigenvectors (in order to form a basis of $\mathbb{R}^2$), we conclude that in this case there is *no* eigenbasis for this linear transformation.

□

Summarizing, in Examples 2.39 and 2.40, we looked at how the matrix of a transformation changes with respect to two different bases that we were given. In Example 2.42, we looked for a particular basis appropriate to the transformation at hand. In Example 2.47, we looked for an eigenbasis with respect to the given transformation. Example 2.42 in this respect fits into the same framework as Example 2.47, but the orthogonal projection has a zero eigenvalue (see the matrix in (2.15)). Example 2.48 illustrates how eigenvectors, in general, need not be orthogonal. In Example 2.49 we see that sometimes an eigenbasis does not exist.

EXERCISE 2.50. Let $\mathcal{B} = \{1, x, x^2, x^3\}$ be the standard basis of the vector space $V := P_{\leq 3}(\mathbb{R})$ of real polynomials of degree at most 3. Moreover, let $\alpha : V \to V$ be the function $\alpha(p(x)) := (x-1)p'(x)$, where $p'(x) := \frac{dp}{dx}(x)$ is the derivative. And let $\widetilde{\mathcal{B}} = \{1, x-1, (x-1)^2, (x-1)^3\}$.

(1) Show that α is a linear transformation.
(2) Show that $\widetilde{\mathcal{B}}$ is an eigenbasis of V for the linear transformation α.
(3) What is the matrix $\widetilde{M}$ representing α with respect to $\widetilde{\mathcal{B}}$?
(4) Let $\beta : V \to \mathbb{R}$ be the linear transformation $f(x) \mapsto f(1)$. Determine the matrices A and $\widetilde{A}$ of the linear transformation β with respect to the bases $\mathcal{B}$ and $\widetilde{\mathcal{B}}$, respectively.
(5) Write the matrix of the change of basis $L_{\mathcal{B}\widetilde{\mathcal{B}}}$ from the (old) basis $\mathcal{B}$ to the (new) basis $\widetilde{\mathcal{B}}$ and write its inverse $L_{\widetilde{\mathcal{B}}\mathcal{B}}$.
(6) Check that $A L_{\mathcal{B}\widetilde{\mathcal{B}}} = \widetilde{A}$.

Multilinear Forms

3

This chapter introduces linear, bilinear and multilinear forms. We explore their definitions, properties, and examples, emphasizing their character as natural generalizations of linear functions and scalar products. These forms allow the seamless transition to covariant tensors.

3.1 Linear Forms

3.1.1 Definition and Examples

Linear forms on a vector space V are defined as linear real-valued functions on V. We will see that linear forms behave very much like vectors, only that they are elements *not* of V, but of a different, yet related, vector space. Whereas we represent regular vectors from V by column vectors once a basis is fixed, we will represent linear forms on V by row vectors. Then the value of a linear form on a specific vector is simply given by the matrix product with the row vector (linear form) on the left and the column vector (actual vector) on the right.

> DEFINITION 3.1. Let V be a vector space. A **linear form** on V is a map $\alpha : V \to \mathbb{R}$ such that for every $a, b \in \mathbb{R}$ and for every $v, w \in V$
>
> $$\alpha(av + bw) = a\alpha(v) + b\alpha(w).$$

Alternative terminologies for "linear form" are **tensor of type** $(0, 1)$, **1-form**, **linear functional** and **covector**.

EXERCISE 3.2. If $V = \mathbb{R}^3$, which of the following are linear forms?

(1) $\alpha(x, y, z) := xy + z$;
(2) $\alpha(x, y, z) := x + y + z + 1$;
(3) $\alpha(x, y, z) := \pi x - \frac{7}{2}z$.

EXERCISE 3.3. If V is the infinite dimensional vector space of continuous functions $f : \mathbb{R} \to \mathbb{R}$, which of the following are linear forms?

(1) $\alpha(f) := f(7) - f(0)$;
(2) $\alpha(f) := \int_0^{33} e^x f(x) dx$;
(3) $\alpha(f) := e^{f(4)}$.

Example 3.4 [Coordinate forms] This is a very important example of linear form. Let $\mathcal{B} = \{b_1, \ldots, b_n\}$ be a basis of V and let $v = v^i b_i \in V$ be a generic vector. Define $\beta^i : V \to \mathbb{R}$ by

$$\boxed{\beta^i(v) := v^i}, \qquad (3.1)$$

that is β^i will extract the i-th coordinate of a vector with respect to the basis $\mathcal{B}$. The linear form β^i is called **coordinate form**. Notice that

$$\beta^i(b_j) = \delta^i_j, \qquad (3.2)$$

since the i-th coordinate of the basis vector b_j with respect to the basis $\mathcal{B}$ is equal to 1 if $i = j$ and 0 otherwise. □

Example 3.5 Let $V = \mathbb{R}^3$ and let $\mathcal{E}$ be its standard basis. The three coordinate forms are defined by

$$\beta^1 \begin{bmatrix} x \\ y \\ z \end{bmatrix} := x, \quad \beta^2 \begin{bmatrix} x \\ y \\ z \end{bmatrix} := y, \quad \beta^3 \begin{bmatrix} x \\ y \\ z \end{bmatrix} := z.$$

□

Example 3.6 Let $V = \mathbb{R}^2$ and let $\mathcal{B} := \{ \underbrace{\begin{bmatrix} 1 \\ 1 \end{bmatrix}}_{b_1}, \underbrace{\begin{bmatrix} 1 \\ -1 \end{bmatrix}}_{b_2} \}$. We want to describe the elements of $\mathcal{B}^* := \{\beta^1, \beta^2\}$, in other words we want to find

$$\beta^1(v) \qquad \text{and} \qquad \beta^2(v)$$

3.1 Linear Forms

for a generic vector $v \in V$.

To this purpose we need to find $[v]_{\mathcal{B}}$. Recall that if $\mathcal{E}$ denotes the standard basis of $\mathbb{R}^2$ and $L := L_{\mathcal{E}\mathcal{B}}$ the matrix of the change of coordinate from $\mathcal{E}$ to $\mathcal{B}$, then

$$[v]_{\mathcal{B}} = L^{-1}[v]_{\mathcal{E}} = L^{-1}\begin{pmatrix} v^1 \\ v^2 \end{pmatrix}.$$

Since

$$L = \begin{bmatrix} 1 & 1 \\ 1 & -1 \end{bmatrix}$$

and hence

$$L^{-1} = \tfrac{1}{2}\begin{bmatrix} 1 & 1 \\ 1 & -1 \end{bmatrix},$$

then

$$[v]_{\mathcal{B}} = \begin{pmatrix} \tfrac{1}{2}(v^1 + v^2) \\ \tfrac{1}{2}(v^1 - v^2) \end{pmatrix}.$$

Thus, according to the definition in Eq. (3.1), we deduce that

$$\beta^1(v) = \tfrac{1}{2}(v^1 + v^2) \quad \text{and} \quad \beta^2(v) = \tfrac{1}{2}(v^1 - v^2).$$

□

3.1.2 Dual Space and Dual Basis

To any vector space V, there corresponds the set of all linear forms on V. This set has itself a natural structure of a vector space.

DEFINITION 3.7. The **dual** (or *dual space*) of a vector space V is

$$V^* := \{\text{all linear forms } \alpha : V \to \mathbb{R}\},$$

equipped with the natural addition and scalar multiplication.

EXERCISE 3.8. Check that V^* is a vector space whose null vector is the linear form identically equal to zero.

Remark 3.9 Just like any function, two linear forms on V are equal if and only if their values are the same when applied to *each* vector in V. However, because of the defining properties of linear forms, to determine whether two linear forms are equal, it is *enough to check that they are equal on each element of a basis of V*. In order to check this, let $\alpha, \alpha' \in V^*$, let $\mathcal{B} = \{b_1, \ldots, b_n\}$ be a basis of V and suppose that we know that

$$\alpha(b_j) = \alpha'(b_j)$$

for all $1 \leq j \leq n$. We now verify that this implies that α and α' are the same when applied to each vector $v \in V$. Let $v = v^j b_j$ be the representation of v with respect to the basis $\mathcal{B}$. Then we have

$$\alpha(v) = \alpha(v^j b_j) = v^j \alpha(b_j) = v^j \alpha'(b_j) = \alpha'(v^j b_j) = \alpha'(v).$$

$\square$

Proposition 3.10 *Let $\mathcal{B} = \{b_1, \ldots, b_n\}$ be a basis of V and $\beta^1, \ldots, \beta^n$ the corresponding coordinate forms. Then $\mathcal{B}^* := \{\beta^1, \ldots, \beta^n\}$ is a basis of V^*. As a consequence*

$$\dim V = \dim V^*.$$

Proof According to Definition 2.22, we need to check that the linear forms in $\mathcal{B}^*$

(1) are linearly independent and
(2) span V^*.

(1) We need to check that the only linear combination of $\beta^1, \ldots, \beta^n$ that yields the zero linear form is the trivial linear combination. Let $c_i \beta^i = 0$ be a linear combination of the β^i. Then for every basis vector b_j, with $j = 1, \ldots, n$,

$$0 = (c_i \beta^i)(b_j) = c_i(\beta^i(b_j)) = c_i \delta^i_j = c_j,$$

thus showing the linear independence.

(2) To check that $\mathcal{B}^*$ spans V we need to verify that any $\alpha \in V^*$ is a linear combination of $\beta^1, \ldots, \beta^n$, that is, that we can find $\alpha_i \in \mathbb{R}$ such that

$$\alpha = \alpha_i \beta^i. \tag{3.3}$$

To find such α_i we apply both sides of Eq. (3.3) to the j-th basis vector b_j, and we obtain

$$\alpha(b_j) = \alpha_i \beta^i(b_j) = \alpha_i \delta^i_j = \alpha_j, \tag{3.4}$$

3.1 Linear Forms

which identifies the coefficients in Eq. (3.3).

By hypothesis α is a linear form and, since V^* is a vector space, also $\alpha(b_i)\beta^i$ is a linear form. Moreover, we have just verified that these two linear form coincide on the basis vectors. By Remark 3.9 the two linear forms are the same and, hence, we have written α as a linear combination of the coordinate forms. This completes the proof that the coordinate forms form a basis of the dual. $\square$

The basis $\mathcal{B}^*$ of V^* is called the **basis of V^* dual to** $\mathcal{B}$. We emphasize that the components (or coordinates) of a linear form α with respect to $\mathcal{B}^*$ are exactly the values of α on the elements of $\mathcal{B}$, as we found in the above proof:

$$\boxed{\alpha_i = \alpha(b_i)}.$$

We build with these the coordinate-vector of α as a *row-vector*:

$$[\alpha]_{\mathcal{B}^*} := \begin{pmatrix} \alpha_1 & \ldots & \alpha_n \end{pmatrix}.$$

Example 3.11 Let $V = P_2(\mathbb{R})$ be the vector space of polynomials of degree ≤ 2, let $\alpha : V \to \mathbb{R}$ be the linear form given by

$$\alpha(p) := p(2) - p'(2) \tag{3.5}$$

and let $\mathcal{B}$ be the basis $\{1, x, x^2\}$ of V. In this example, we want to:

(1) find the components of α with respect to $\mathcal{B}^*$;
(2) describe the basis $\mathcal{B}^* = \{\beta^1, \beta^2, \beta^3\}$;

(1) Since

$$\alpha_1 = \alpha(b_1) = \alpha(1) = 1 - 0 = 1$$
$$\alpha_2 = \alpha(b_2) = \alpha(x) = 2 - 1 = 1$$
$$\alpha_3 = \alpha(b_3) = \alpha(x^2) = 4 - 4 = 0,$$

then

$$[\alpha]_{\mathcal{B}^*} = \begin{pmatrix} 1 & 1 & 0 \end{pmatrix}. \tag{3.6}$$

(2) The generic element $p(x) \in P_2(\mathbb{R})$ written as combination of basis elements $1, x$ and x^2 is

$$p(x) = a + bx + cx^2.$$

Hence $\mathcal{B}^* = \{\beta^1, \beta^2, \beta^3\}$, is given by

$$\beta^1(a + bx + cx^2) = a$$
$$\beta^2(a + bx + cx^2) = b \qquad (3.7)$$
$$\beta^3(a + bx + cx^2) = c.$$

□

Remark 3.12 Note that we have to be careful when referring to a "dual basis" of V^*, as for every basis $\mathcal{B}$ of V there is going to be a basis $\mathcal{B}^*$ of V^* dual to the basis $\mathcal{B}$. In the next section we are going to see how a dual basis transforms with a change of basis.

3.1.3 Covariance of Linear Forms

We want to examine how a linear form $\alpha : V \to \mathbb{R}$ behaves with respect to a change a basis in V. To this purpose, let

$$\mathcal{B} = \{b_1, \ldots, b_n\} \quad \text{and} \quad \widetilde{\mathcal{B}} := \{\tilde{b}_1, \ldots, \tilde{b}_n\}$$

be two bases of V and let

$$\mathcal{B}^* := \{\beta^1, \ldots, \beta^n\} \quad \text{and} \quad \widetilde{\mathcal{B}}^* := \{\tilde{\beta}^1, \ldots, \tilde{\beta}^n\}$$

be the corresponding dual bases. Let

$$[\alpha]_{\mathcal{B}^*} = (\alpha_1 \ldots \alpha_n) \quad \text{and} \quad [\alpha]_{\widetilde{\mathcal{B}}^*} = (\tilde{\alpha}_1 \ldots \tilde{\alpha}_n)$$

be the coordinate vectors of α with respect to $\mathcal{B}^*$ and $\widetilde{\mathcal{B}}^*$, that is

$$\alpha(b_i) = \alpha_i \quad \text{and} \quad \alpha(\tilde{b}_i) = \tilde{\alpha}_i.$$

Let $L := L_{\mathcal{B}\widetilde{\mathcal{B}}}$ be the matrix of the change of basis satisfying Eq. (2.4)

$$\tilde{b}_j = L^i_j b_i.$$

Then we have

$$\tilde{\alpha}_j = \alpha(\tilde{b}_j) = \alpha(L^i_j b_i) = L^i_j \alpha(b_i) = L^i_j \alpha_i = \alpha_i L^i_j, \qquad (3.8)$$

3.1 Linear Forms

so that

$$\boxed{\tilde{\alpha}_j = \alpha_i L^i_j}. \tag{3.9}$$

EXERCISE 3.13. Verify that Eq. (3.9) is equivalent to saying that

$$[\alpha]_{\widetilde{\mathcal{B}}^*} = [\alpha]_{\mathcal{B}^*} L. \tag{3.10}$$

Note that we have exchanged the order of α_i and L^i_j in the last equality of Eq. (3.8) to respect the order in which the matrix multiplication in Eq. (3.10) has to be performed. This was possible because both α_i and L^i_j are real numbers.

We say that a linear form α is **covariant** because its components change by L when the basis changes by L. A linear form α is hence a **covariant tensor** or a **tensor of type** $(0, 1)$.

Example 3.14 We continue with Example 3.11. We consider the bases as in Example 2.41, that is

$$\mathcal{B} := \{1, x, x^2\} \quad \text{and} \quad \widetilde{\mathcal{B}} := \{x, x - 1, x^2 - 1\}$$

and the linear form $\alpha : V \to \mathbb{R}$ as in Eq. (3.5). We will:

(1) find the components of α with respect to $\mathcal{B}^*$;
(2) describe the basis $\mathcal{B}^* = \{\beta^1, \beta^2, \beta^3\}$;
(3) find the components of α with respect to $\widetilde{\mathcal{B}}^*$;
(4) describe the basis $\widetilde{\mathcal{B}}^* = \{\tilde{\beta}^1, \tilde{\beta}^2, \tilde{\beta}^3\}$;
(5) find the matrix of change of basis $L := L_{\mathcal{B}\widetilde{\mathcal{B}}}$ and compute $\Lambda = L^{-1}$;
(6) check the covariance of α;
(7) check the contravariance of $\mathcal{B}^*$.

(1) This is done in Eq. (3.6).
(2) This is done in Eq. (3.7).
(3) We proceed as in Eq. (3.6). Namely,

$$\tilde{\alpha}_1 = \alpha(\tilde{b}_1) = \alpha(x) = 2 - 1 = 1$$
$$\tilde{\alpha}_2 = \alpha(\tilde{b}_2) = \alpha(x - 1) = 1 - 1 = 0$$
$$\tilde{\alpha}_3 = \alpha(\tilde{b}_3) = \alpha(x^2 - 1) = 3 - 4 = -1,$$

so that

$$[\alpha]_{\widetilde{\mathcal{B}}^*} = \begin{pmatrix} 1 & 0 & -1 \end{pmatrix}.$$

(4) Since $\tilde{\beta}^i(v) = \tilde{v}^i$, to proceed as in Eq. (3.7) we first need to write the generic polynomial $p(x) = a + bx + cx^2$ as a linear combination of elements in $\widetilde{\mathcal{B}}$, namely we need to find $\tilde{a}, \tilde{b}$ and $\tilde{c}$ such that

$$p(x) = a + bx + cx^2 = \tilde{a}x + \tilde{b}(x-1) + \tilde{c}(x^2 - 1).$$

By multiplying and collecting the terms, we obtain that

$$\begin{cases} -\tilde{b} - \tilde{c} = a \\ \tilde{a} + \tilde{b} = b \\ \tilde{c} = c \end{cases} \quad \text{that is} \quad \begin{cases} \tilde{a} = a + b + c \\ \tilde{b} = -a - c \\ \tilde{c} = c. \end{cases}$$

Hence

$$p(x) = a + bx + cx^2 = (a+b+c)x + (-a-c)(x-1) + c(x^2 - 1),$$

so that it follows that

$$\tilde{\beta}^1(p(x)) = a + b + c$$
$$\tilde{\beta}^2(p(x)) = -a - c$$
$$\tilde{\beta}^3(p(x)) = c,$$

(5) The matrix of the change of basis is given by

$$L := L_{\mathcal{B}\widetilde{\mathcal{B}}} = \begin{bmatrix} 0 & -1 & -1 \\ 1 & 1 & 0 \\ 0 & 0 & 1 \end{bmatrix},$$

since for example $\tilde{b}_3$ can be written as a linear combination with respect to $\mathcal{B}$ as $\tilde{b}_3 = x^2 - 1 = -1b_1 + 0b_2 + 1b_3$, and these coordinates $-1, 0, 1$ build the third column of L.

To compute $\Lambda = L^{-1}$ we can use the Gauss–Jordan elimination process

$$\left[\begin{array}{ccc|ccc} 0 & -1 & -1 & 1 & 0 & 0 \\ 1 & 1 & 0 & 0 & 1 & 0 \\ 0 & 0 & 1 & 0 & 0 & 1 \end{array}\right] \quad \longleftrightarrow \quad \ldots \quad \longleftrightarrow \quad \left[\begin{array}{ccc|ccc} 1 & 0 & 0 & 1 & 1 & 1 \\ 0 & 1 & 0 & -1 & 0 & -1 \\ 0 & 0 & 1 & 0 & 0 & 1 \end{array}\right]$$

Therefore, we have

$$\Lambda = \begin{bmatrix} 1 & 1 & 1 \\ -1 & 0 & -1 \\ 0 & 0 & 1 \end{bmatrix}.$$

3.1 Linear Forms

(6) The linear form α is indeed *covariant*, since

$$(\alpha_1 \; \alpha_2 \; \alpha_3) L = (1 \; 1 \; 0) \begin{bmatrix} 0 & -1 & -1 \\ 1 & 1 & 0 \\ 0 & 0 & 1 \end{bmatrix} = (1 \; 0 \; -1) = (\tilde{\alpha}_1 \; \tilde{\alpha}_2 \; \tilde{\alpha}_3).$$

(7) The dual basis $\mathcal{B}^*$ is *contravariant*, since

$$\begin{pmatrix} \tilde{\beta}^1 \\ \tilde{\beta}^2 \\ \tilde{\beta}^3 \end{pmatrix} = \Lambda \begin{pmatrix} \beta^1 \\ \beta^2 \\ \beta^3 \end{pmatrix},$$

as it can be verified by evaluating both sides on an arbitrary vector $p(x) = a + bx + cx^2$:

$$\Lambda \begin{pmatrix} \beta^1(p) \\ \beta^2(p) \\ \beta^3(p) \end{pmatrix} = \begin{bmatrix} 1 & 1 & 1 \\ -1 & 0 & -1 \\ 0 & 0 & 1 \end{bmatrix} \begin{pmatrix} a \\ b \\ c \end{pmatrix} = \begin{pmatrix} a+b+c \\ -a-c \\ c \end{pmatrix} = \begin{pmatrix} \tilde{\beta}^1(p) \\ \tilde{\beta}^2(p) \\ \tilde{\beta}^3(p) \end{pmatrix}.$$

$\square$

3.1.4 Contravariance of Dual Bases

In fact, statement (7) in Example 3.11 holds in general, namely:

Proposition 3.15 *Dual bases are **contravariant**.*

Proof We will check that when bases $\mathcal{B}$ and $\tilde{\mathcal{B}}$ are related by

$$\tilde{b}_j = L^i_j b_i$$

the corresponding dual bases $\mathcal{B}^*$ and $\tilde{\mathcal{B}}^*$ of V^* are related by

$$\boxed{\tilde{\beta}^j = \Lambda^j_i \beta^i}. \tag{3.11}$$

It is enough to check that the $\Lambda^j_i \beta^i$ are *dual* to the $\tilde{b}_j$. But since $\Lambda L = I$, we have

$$(\Lambda^k_\ell \beta^\ell)(\tilde{b}_j) = (\Lambda^k_\ell \beta^\ell)(L^i_j b_i) = \Lambda^k_\ell L^i_j \beta^\ell(b_i) = \Lambda^k_\ell L^i_j \delta^\ell_i = \Lambda^k_i L^i_j = \delta^k_j = \beta^j(\tilde{b}_j).$$

$\square$

Table 3.1 Duality

V real vector space	$V^* = \{$linear forms $\alpha : V \to \mathbb{R}\}$
with $\dim V = n$	dual vector space to V
$\mathcal{B} = \{b_1, \ldots, b_n\}$	$\mathcal{B}^* = \{\beta^1, \ldots, \beta^n\}$
basis of V	dual basis of V^* w.r.t. $\mathcal{B}$
$\widetilde{\mathcal{B}} := \{\tilde{b}_1, \ldots, \tilde{b}_n\}$	$\widetilde{\mathcal{B}}^* = \{\tilde{\beta}^1, \ldots, \tilde{\beta}^n\}$
another basis of V	dual basis of V^* w.r.t. $\widetilde{\mathcal{B}}$
$L := L_{\mathcal{B}\widetilde{\mathcal{B}}}$ =matrix of the change	$\Lambda = L^{-1}$ =matrix of the change
of basis from $\mathcal{B}$ to $\widetilde{\mathcal{B}}$	of basis from $\widetilde{\mathcal{B}}$ to $\mathcal{B}$
Then we have $\tilde{b}_j = L_j^i b_i$ or	Then we have $\tilde{\beta}^i = \Lambda_j^i \beta^j$ or
$\boxed{\begin{pmatrix} \tilde{b}_1 & \ldots & \tilde{b}_n \end{pmatrix} = \begin{pmatrix} b_1 & \ldots & b_n \end{pmatrix} L}$	$\boxed{\begin{pmatrix} \tilde{\beta}^1 \\ \vdots \\ \tilde{\beta}^n \end{pmatrix} = L^{-1} \begin{pmatrix} \beta^1 \\ \vdots \\ \beta^n \end{pmatrix}}$
covariance of a basis	contravariance of the dual basis
If v is any vector in V,	If α is any linear form in V^*,
then $v = v^i b_i = \tilde{v}^i \tilde{b}_i$, where	then $\alpha = \alpha_j \beta^j = \tilde{\alpha}_j \tilde{\beta}^j$, where
$\tilde{v}^i = \Lambda_j^i v^j$ i.e., $[v]_{\widetilde{\mathcal{B}}} = L^{-1}[v]_{\mathcal{B}}$ or	$\tilde{\alpha}_j = L_j^i \alpha_i$ i.e., $[\alpha]_{\widetilde{\mathcal{B}}^*} = [\alpha]_{\mathcal{B}^*} L$ or
$\boxed{\begin{pmatrix} \tilde{v}^1 \\ \vdots \\ \tilde{v}^n \end{pmatrix} = L^{-1} \begin{pmatrix} v^1 \\ \vdots \\ v^n \end{pmatrix}}$	$\boxed{\begin{pmatrix} \tilde{\alpha}_1 & \ldots & \tilde{\alpha}_n \end{pmatrix} = \begin{pmatrix} \alpha_1 & \ldots & \alpha_n \end{pmatrix} L}$
contravariance of the coordinate vectors	covariance of linear forms components
$\to$ vectors are $(1, 0)$-tensors	$\to$ linear forms are $(0, 1)$-tensors

Table 3.2 Covariance vs. contravariance

	Covariance of a tensor	Contravariance of a tensor
Is denoted by	Lower indices	Upper indices
Coordinate-vectors are indicated as	Row vectors	Column vectors
The tensor transforms w.r.t. a change of basis from $\mathcal{B}$ to $\widetilde{\mathcal{B}}$ by multiplication with	L on the right	L^{-1} on the left
(For later use) A tensor of type (p, q) has	Covariant order q	Contravariant order p

Table 3.1 contains a summary of the properties that bases and dual bases, coordinate vectors and components of linear forms satisfy with respect to a change of basis. Table 3.2 summarizes the characteristics of covariance and contravariance.

3.2 Bilinear Forms

3.2.1 Definition and Examples

DEFINITION 3.16. A **bilinear form** on V is a function $\varphi : V \times V \to \mathbb{R}$ that is linear in each variable, that is

$$\varphi(u, \lambda v + \mu w) = \lambda \varphi(u, v) + \mu \varphi(u, w)$$
$$\varphi(\lambda v + \mu w, u) = \lambda \varphi(v, u) + \mu \varphi(w, u),$$

for every $\lambda, \mu \in \mathbb{R}$ and for every $u, v, w \in V$.

Examples 3.17 Let $V = \mathbb{R}^n$.

(1) If $v, w \in \mathbb{R}^n$, the **dot product** (or **scalar product**) defined as

$$\varphi(v, w) = v \cdot w = v^j w^j \delta_{ij},$$

(see also Exercise 4.2).

(2) Let $n = 3$. Choose a vector $u \in \mathbb{R}^3$ and for any two vectors $v, w \in \mathbb{R}^3$, denote by $v \times w$ their **cross product**. The **scalar triple product**[1]

---

[1] Recall that the determinant of a 3×3 matrix is given by

$$\det \begin{bmatrix} a_{11} & a_{12} & a_{13} \\ a_{21} & a_{22} & a_{23} \\ a_{31} & a_{32} & a_{33} \end{bmatrix} = a_{11}a_{22}a_{33} - a_{11}a_{23}a_{32} + a_{12}a_{23}a_{31}$$

$$- a_{12}a_{21}a_{33} + a_{13}a_{21}a_{32} - a_{13}a_{22}a_{31}$$

$$= \sum_{\sigma \in S_3} \text{sign}(\sigma) a_{1\sigma(1)} a_{2\sigma(2)} a_{3\sigma(3)},$$

where

$$\sigma = (\sigma(1), \sigma(2), \sigma(3)) \in S_3 := \{\text{permutations of 3 elements}\}$$
$$= \{(1, 2, 3), (1, 3, 2), (2, 3, 1), (2, 1, 3), (3, 1, 2), (3, 2, 1)\},$$

and the corresponding signs flip each time two elements get swapped:

$$\text{sign}(1, 2, 3) = 1, \quad \text{sign}(1, 3, 2) = -1, \quad \text{sign}(3, 1, 2) = 1,$$
$$\text{sign}(3, 2, 1) = -1, \quad \text{sign}(2, 3, 1) = 1, \quad \text{sign}(2, 1, 3) = -1.$$

An **even permutation** is a permutation σ with $\text{sign}(\sigma) = 1$; an **odd permutation** is a permutation σ with $\text{sign}(\sigma) = -1$.

$$\varphi^u(v, w) := u \cdot (v \times w) = \det \begin{bmatrix} u \\ v \\ w \end{bmatrix} \quad (3.12)$$

is a bilinear form in v and w, where $\begin{bmatrix} u \\ v \\ w \end{bmatrix}$ denotes the matrix with rows u, v and w. The quantity $\varphi^u(v, w)$ calculates the signed volume of the parallelepiped spanned by u, v, w: the sign of $\varphi_u(v, w)$ depends on the orientation of the triple u, v, w.

Since the cross product is defined *only* in $\mathbb{R}^3$, in contrast with the scalar product, the scalar triple product cannot be defined in $\mathbb{R}^n$ with $n \neq 3$ (though there is a formula for an n dimensional paralleldiped involving some "generalization" of it). □

EXERCISE 3.18. Verify the equality in Eq. (3.12) using the Laplace expansion formula for the determinant of a 3×3 matrix with respect to the first line. Recall that this reads

$$\det \begin{bmatrix} u^1 & u^2 & u^3 \\ v^1 & v^2 & v^3 \\ w^1 & w^2 & w^3 \end{bmatrix} = u^1 \det \begin{bmatrix} v^2 & v^3 \\ w^2 & w^3 \end{bmatrix} - u^2 \det \begin{bmatrix} v^1 & v^3 \\ w^1 & w^3 \end{bmatrix} + u^3 \det \begin{bmatrix} v^1 & v^2 \\ w^1 & w^2 \end{bmatrix}$$

Examples 3.19 Let $V = P_2(\mathbb{R})$.

(1) Let $p, q \in P_2(\mathbb{R})$. The function $\varphi(p, q) := p(\pi)q(33)$ is a bilinear form.
(2) Likewise,

$$\varphi(p, q) := p'(0)q(4) - 5p'(3)q''(\tfrac{1}{2})$$

is a bilinear form. □

EXERCISE 3.20. Are the following functions bilinear forms?

(1) $V = \mathbb{R}^2$ and $\varphi(u, v) := \det \begin{bmatrix} u \\ v \end{bmatrix}$;
(2) $V = P_2(\mathbb{R})$ and $\varphi(p, q) := \int_0^1 p(x)q(x)dx$;

3.2 Bilinear Forms

(3) $V = M_{2\times 2}(\mathbb{R})$, the space of real 2×2 matrices, and $\varphi(L, M) := L_1^1 \operatorname{tr} M$, where L_1^1 it the (1,1)-entry of L and $\operatorname{tr} M$ is the trace of M;

(4) $V = \mathbb{R}^3$ and $\varphi(v, w) := v \times w$;

(5) $V = \mathbb{R}^2$ and $\varphi(v, w)$ is the area of the parallelogram spanned by v and w.

(6) $V = M_{n\times n}(\mathbb{R})$, the space of real $n \times n$ matrices with $n > 1$, and $\varphi(L, M) := \operatorname{tr} L \det M$, where $\operatorname{tr} L$ is the trace of L and $\det M$ is the determinant of M.

Remark 3.21 We need to be careful about the following possible confusion. A bilinear form on V is a function on $V \times V$ that is linear in each variable *separately*. But $V \times V$ is also a vector space and one might wonder whether a bilinear form on V is also a linear form on the vector space $V \times V$. But this is not the case. For example, consider the case in which $V = \mathbb{R}$, so that $V \times V = \mathbb{R}^2$ and let $\varphi : \mathbb{R} \times \mathbb{R} \to \mathbb{R}$ be a function:

(1) If $\varphi(x, y) := 2x - y$, then φ is not a *bilinear form on* $\mathbb{R}$, but is a *linear form* on $(x, y) \in \mathbb{R}^2$;

(2) If $\varphi(x, y) := 2xy$, then φ is a *bilinear form on* $\mathbb{R}$ (hence *linear* in $x \in \mathbb{R}$ and *linear* in $y \in \mathbb{R}$), but it is *not* a linear form on $\mathbb{R}^2$, as it is *not linear* in $(x, y) \in \mathbb{R}^2$.

So a *bi*linear form is not a form that it is "twice as linear" as a linear form, but a form that is defined on the product of twice the vector space. □

EXERCISE 3.22. Verify the above assertions in Remark 3.21 to make sure the distinction is clear.

3.2.2 Tensor Product of Two Linear Forms on V

Let $\alpha, \beta \in V^*$ be two linear forms, $\alpha, \beta : V \to \mathbb{R}$, and define $\varphi : V \times V \to \mathbb{R}$, by

$$\varphi(v, w) := \alpha(v)\beta(w).$$

Then φ is bilinear, is called the **tensor product** of the linear forms α and β and is denoted by

$$\varphi = \alpha \otimes \beta.$$

Note 3.23 In general $\alpha \otimes \beta \neq \beta \otimes \alpha$, as there could be vectors v and w such that $\alpha(v)\beta(w) \neq \beta(v)\alpha(w)$.

Example 3.24 Let $V = P_2(\mathbb{R})$, let $\alpha(p) = p(2) - p'(2)$ and $\beta(p) = \int_3^4 p(x)dx$ be two linear forms. Then

$$(\alpha \otimes \beta)(p, q) = (p(2) - p'(2)) \int_3^4 q(x)dx$$

is a bilinear form. □

Remark 3.25 The bilinear form $\varphi : \mathbb{R} \times \mathbb{R} \to \mathbb{R}$ defined by the formula $\varphi(x, y) := 2xy$ is the tensor product of two linear forms on $\mathbb{R}$, for instance, $\varphi(x, y) = (\alpha \otimes \alpha)(x, y)$ where $\alpha : \mathbb{R} \to \mathbb{R}$ is the linear form given by $\alpha(x) := \sqrt{2}x$.

On the other hand, not every bilinear form is simply the tensor product of two linear forms. As we will see below, the first such examples are found for bilinear forms on vector spaces of dimension at least 2. □

3.2.3 A Basis for Bilinear Forms

Let

$$\mathrm{Bil}(V \times V, \mathbb{R}) := \{\text{all bilinear forms } \varphi : V \times V \to \mathbb{R}\}.$$

EXERCISE 3.26. Check that $\mathrm{Bil}(V \times V, \mathbb{R})$ is a vector space with the zero element equal to the bilinear form identically equal to zero.

Hint: It is enough to check that if $\varphi, \psi \in \mathrm{Bil}(V \times V, \mathbb{R})$, and $\lambda, \mu \in \mathbb{R}$, then $\lambda \varphi + \mu \psi \in \mathrm{Bil}(V \times V, \mathbb{R})$. Why? Recall Example 2.3(3) in Sect. 2.1.1 and Exercise 3.8 in Sect. 3.1.2.

Assuming Exercise 3.26, we are going to find a basis of $\mathrm{Bil}(V \times V, \mathbb{R})$ and determine its dimension. Let $\mathcal{B} = \{b_1, \ldots, b_n\}$ be a basis of V and let $\mathcal{B}^* = \{\beta^1, \ldots, \beta^n\}$ be the dual basis of V^* (that is $\beta^i(b_j) = \delta^i_j$).

Proposition 3.27 *The bilinear forms $\beta^i \otimes \beta^j$, $i, j = 1, \ldots, n$ form a basis of $\mathrm{Bil}(V \times V, \mathbb{R})$. As a consequence,* $\dim \mathrm{Bil}(V \times V, \mathbb{R}) = n^2$.

Notation We denote

$$\boxed{\mathrm{Bil}(V \times V, \mathbb{R}) = V^* \otimes V^*}$$

and call this vector space the **tensor product** of V^* and V^*. A justification for this notation will appear in Sect. 5.4.2.

Remark 3.28 Just as it is for linear forms, to verify that two bilinear forms on V are the same it is enough to verify that they are the same on every pair of elements

3.2 Bilinear Forms

of a basis of V. So let φ, ψ be two bilinear forms, let $\mathcal{B} = \{b_1, \ldots, b_n\}$ be a basis of V, and assume that

$$\varphi(b_i, b_j) = \psi(b_i, b_j)$$

for all $1 \leq i, j, \leq n$. Let $v = v^i b_i, w = w^j b_j \in V$ be arbitrary vectors. We now verify that $\varphi(v, w) = \psi(v, w)$. Because of the linearity in each variable, we have

$$\varphi(v, w) = \varphi(v^i b_i, w^j b_j) = v^i w^j \varphi(b_i, b_j) = v^i w^j \psi(b_i, b_j)$$
$$= \psi(v^i b_i, w^j b_j) = \psi(v, w).$$

□

Proof of Proposition 3.27 The proof will be similar to the one of Proposition 3.10 for linear forms. We first check that the set of bilinear forms $\{\beta^i \otimes \beta^j, i, j = 1, \ldots, n\}$ consists of linearly independent vectors, then that it spans $\text{Bil}(V \times V, \mathbb{R})$.

For the linear independence we need to check that the only linear combination of the $\beta^i \otimes \beta^j$ that gives the zero bilinear form is the trivial linear combination. Let $c_{ij} \beta^i \otimes \beta^j = 0$ be a linear combination of the $\beta^i \otimes \beta^j$. Then for all pairs of basis vectors (b_k, b_ℓ), with $k, \ell = 1, \ldots, n$, we have

$$0 = c_{ij} \beta^i \otimes \beta^j (b_k, b_\ell) = c_{ij} \delta^i_k \delta^j_\ell = c_{k\ell},$$

thus showing the linear independence.

To check that $\text{span}\{\beta^i \otimes \beta^j, i, j = 1, \ldots, n\} = \text{Bil}(V \times V, \mathbb{R})$, we need to check that if $\varphi \in \text{Bil}(V \times V, \mathbb{R})$, there exists $B_{ij} \in \mathbb{R}$ such that

$$\varphi = B_{ij} \beta^i \otimes \beta^j.$$

Because of Eq. (3.2) in Sect. 3.1.1, we obtain

$$\varphi(b_k, b_\ell) = B_{ij} \beta^i(b_k) \beta^j(b_\ell) = B_{ij} \delta^i_k \delta^j_\ell = B_{k\ell},$$

for every pair $(b_k, b_\ell) \in V \times V$. Hence, we set $B_{k\ell} := \varphi(b_k, b_\ell)$. Now both φ and $\varphi(b_k, b_\ell) \beta^i \otimes \beta^j$ are bilinear forms and they coincide on $\mathcal{B} \times \mathcal{B}$. Because of the above Remark 3.28, the two bilinear forms coincide. □

Example 3.29 We continue with the study of the *scalar triple product*

$$\varphi^u : \mathbb{R}^3 \times \mathbb{R}^3 \to \mathbb{R},$$

that was defined in Example 3.17 for a fixed given vector $u = \begin{bmatrix} u^1 \\ u^2 \\ u^3 \end{bmatrix}$. We now want to find the components B_{ij} of φ^u with respect to the standard basis of $\mathbb{R}^3$.

Recall the cross product in $\mathbb{R}^3$ is defined on the elements of the standard basis by

$$e_i \times e_j := \begin{cases} 0 & \text{if } i = j \\ e_k & \text{if } (i, j, k) \text{ is a cyclic permutation of } (1, 2, 3) \\ -e_k & \text{if } (i, j, k) \text{ is a non-cyclic permutation of } (1, 2, 3), \end{cases}$$

that is

$$\text{cyclic} \begin{cases} e_1 \times e_2 = e_3 \\ e_2 \times e_3 = e_1 \\ e_3 \times e_1 = e_2 \end{cases} \text{ and } \text{non-cyclic} \begin{cases} e_2 \times e_1 = -e_3 \\ e_3 \times e_2 = -e_1 \\ e_1 \times e_3 = -e_2 \end{cases}$$

Since $u \cdot e_k = u^k$, then

$$B_{ij} = \varphi_u(e_i, e_j) = u \cdot (e_i \times e_j)$$

$$= \begin{cases} 0 & \text{if } i = j \\ u^k & \text{if } (i, j, k) \text{ is a cyclic permutation of } (1, 2, 3) \\ -u^k & \text{if } (i, j, k) \text{ is a non-cyclic permutation of } (1, 2, 3) \end{cases}$$

Thus

$$B_{12} = u^3 = -B_{21}$$
$$B_{31} = u^2 = -B_{13}$$
$$B_{23} = u^1 = -B_{32}$$
$$B_{11} = B_{22} = B_{33} = 0 \quad \text{(that is, the diagonal components are zero)},$$

which can be written as a matrix

$$B = \begin{bmatrix} 0 & u^3 & -u^2 \\ -u^3 & 0 & u^1 \\ u^2 & -u^1 & 0 \end{bmatrix}.$$

The components B_{ij} of B are the components of this bilinear form with respect to the basis $\beta^i \otimes \beta^j$ ($i, j = 1, \ldots, n$), where $\beta^i(e_k) = \delta^i_k$. Hence, we can write

$$\varphi^u = B_{ij} \beta^i \otimes \beta^j = u_1 \left(\beta^2 \otimes \beta^3 - \beta^3 \otimes \beta^2 \right)$$
$$+ u_2 \left(\beta^3 \otimes \beta^1 - \beta^1 \otimes \beta^3 \right) + u_3 \left(\beta^1 \otimes \beta^2 - \beta^2 \otimes \beta^1 \right).$$

$\square$

3.2.4 Covariance of Bilinear Forms

We have seen that, once we choose a basis $\mathcal{B} = \{b_1, \ldots, b_n\}$ of V, we automatically have a basis $\mathcal{B}^* = \{\beta^1, \ldots, \beta^n\}$ of V^* and a basis $\{\beta^i \otimes \beta^j, i, j = 1, \ldots, n\}$ of $V^* \otimes V^*$. This implies, that any bilinear form $\varphi : V \times V \to \mathbb{R}$ can be represented by its components

$$B_{ij} = \varphi(b_i, b_j), \qquad (3.13)$$

in the sense that

$$\varphi = B_{ij} \beta^i \otimes \beta^j.$$

Moreover, these components can be arranged in a matrix[2]

$$B := \begin{bmatrix} B_{11} & \ldots & B_{1n} \\ \vdots & & \vdots \\ B_{n1} & \ldots & B_{nn} \end{bmatrix}$$

called the **matrix of the bilinear form** φ with respect to the chosen basis $\mathcal{B}$. The natural question of course is: how does the matrix B change when we choose a different basis of V?

So, let us choose a different basis $\widetilde{\mathcal{B}} := \{\tilde{b}_1, \ldots, \tilde{b}_n\}$ and corresponding bases $\widetilde{\mathcal{B}}^* = \{\tilde{\beta}^1, \ldots, \tilde{\beta}^n\}$ of V^* and $\{\tilde{\beta}^i \otimes \tilde{\beta}^j, i, j = 1, \ldots, n\}$ of $V^* \otimes V^*$, with respect to which φ will be represented by a matrix $\widetilde{B}$, whose entries are $\widetilde{B}_{ij} = \varphi(\tilde{b}_i, \tilde{b}_j)$.

To see the relation between B and $\widetilde{B}$, due to the change of basis from $\mathcal{B}$ to $\widetilde{\mathcal{B}}$, we start with the matrix of the change of basis $L := L_{\mathcal{B}\widetilde{\mathcal{B}}}$, according to which

$$\tilde{b}_j = L^i_j b_i . \qquad (3.14)$$

Then

$$\widetilde{B}_{ij} = \varphi(\tilde{b}_i, \tilde{b}_j) = \varphi(L^k_i b_k, L^\ell_j b_\ell) = L^k_i L^\ell_j \varphi(b_k, b_\ell) = L^k_i L^\ell_j B_{k\ell},$$

where the first and the last equality follow from Eq. (3.13), the second from Eq. (3.14) (after having renamed the dummy indices to avoid conflicts) and the remaining one from the bilinearity of σ. We conclude that

$$\widetilde{B}_{ij} = L^k_i L^\ell_j B_{k\ell}.$$

---

[2] Contrary to the matrix that gives the change of coordinates between two basis of the vector space, here we have only lower indices. This is not by chance and reflects the type of tensor a bilinear form is.

EXERCISE 3.30. Show that the formula of the transformation of the component of a bilinear form in terms of the matrices of the change of coordinates is

$$\boxed{\widetilde{B} = {}^{t}LBL}, \tag{3.15}$$

where ${}^t L$ denotes the transpose of the matrix L.

We hence say that a bilinear form φ is a **covariant 2-tensor** or a **tensor of type** $(0, 2)$.

3.3 Multilinear Forms

3.3.1 Definition, Basis and Covariance

We saw in Sect. 3.1.3 that linear forms are covariant 1-tensors—or tensors of type $(0, 1)$—and in Sect. 3.2.4 that bilinear forms are covariant 2-tensors—or tensors of type $(0, 2)$.

Analogously to what was done until now, one can define **trilinear forms** on V, that is functions $T : V \times V \times V \to \mathbb{R}$ that are linear with respect to each of the three arguments. The space of trilinear forms on V is denoted

$$V^* \otimes V^* \otimes V^*,$$

has basis

$$\{\beta^j \otimes \beta^j \otimes \beta^k, \ i, j, k = 1, \ldots, n\}$$

and, hence, has dimension n^3. The tensor product $\otimes$ is defined as above.

Since the components of a trilinear form $T : V \times V \times V \to \mathbb{R}$ satisfy the following transformation with respect to a change of basis

$$\widetilde{T}_{ijk} = L_i^\ell L_j^p L_k^q T_{\ell p q},$$

a trilinear form is a **covariant 3-tensor** or a **tensor of type** $(0, 3)$.

Of course, there is nothing special about $k = 1, 2$ or 3:

DEFINITION 3.31. A k-**linear form** or **multilinear form of order** k on V is a function $f : V \times \cdots \times V \to \mathbb{R}$ from k-copies of V into $\mathbb{R}$, that is linear in each of its arguments.

3.3 Multilinear Forms

A k-linear form is a **covariant k-tensor** (or a **covariant tensor of order k** or a **tensor of type** $(0, k)$). The vectors space of k-linear forms on V, denoted

$$\underbrace{V^* \otimes \cdots \otimes V^*}_{k \text{ factors}},$$

has basis

$$\beta^{i_1} \otimes \beta^{i_2} \otimes \cdots \otimes \beta^{i_k}, \quad i_1, \ldots, i_k = 1, \ldots, n$$

and, hence, $\dim(V^* \otimes \cdots \otimes V^*) = n^k$.

3.3.2 Examples of Multilinear Forms

Example 3.32 We once more address the *scalar triple product*, discussed in Examples 3.17 and 3.29. This time we want to find the components B_{ij} of φ^u with respect to the (non-standard) basis

$$\widetilde{\mathcal{B}} := \{ \underbrace{\begin{bmatrix} 0 \\ 1 \\ 0 \end{bmatrix}}_{\tilde{b}_1}, \underbrace{\begin{bmatrix} 1 \\ 0 \\ 1 \end{bmatrix}}_{\tilde{b}_2}, \underbrace{\begin{bmatrix} 0 \\ 0 \\ 1 \end{bmatrix}}_{\tilde{b}_3} \}.$$

The matrix of the change of coordinates from the standard basis to $\widetilde{\mathcal{B}}$ is

$$L = \begin{bmatrix} 0 & 1 & 0 \\ 1 & 0 & 0 \\ 0 & 1 & 1 \end{bmatrix},$$

so that

$$\widetilde{B} = \underbrace{\begin{bmatrix} 0 & 1 & 0 \\ 1 & 0 & 1 \\ 0 & 0 & 1 \end{bmatrix}}_{{}^tL} \underbrace{\begin{bmatrix} 0 & u^3 & -u^2 \\ -u^3 & 0 & u^1 \\ u^2 & -u^1 & 0 \end{bmatrix}}_{B} \underbrace{\begin{bmatrix} 0 & 1 & 0 \\ 1 & 0 & 0 \\ 0 & 1 & 1 \end{bmatrix}}_{L}$$

$$= \underbrace{\begin{bmatrix} 0 & 1 & 0 \\ 1 & 0 & 1 \\ 0 & 0 & 1 \end{bmatrix}}_{{}^tL} \underbrace{\begin{bmatrix} u^3 & -u^2 & -u^2 \\ 0 & u^1 - u^3 & u^1 \\ -u^1 & u^2 & 0 \end{bmatrix}}_{BL} = \begin{bmatrix} 0 & u^1 - u^3 & u^1 \\ u^3 - u^1 & 0 & -u^2 \\ -u^1 & u^2 & 0 \end{bmatrix}.$$

It is easy to check that $\widetilde{B}$ is antisymmetric just like B is, and to check that the components of $\widetilde{B}$ are correct by using the formula for φ:

$$\widetilde{B}_{12} = \varphi(\tilde{b}_1, \tilde{b}_2) = u \cdot (e_2 \times (e_1 + e_3)) = u^1 - u^3$$

$$\widetilde{B}_{13} = \varphi(\tilde{b}_1, \tilde{b}_3) = u \cdot ((e_2) \times e_3) = u^1$$

$$\widetilde{B}_{23} = \varphi(\tilde{b}_2, \tilde{b}_3) = u \cdot ((e_1 + e_3) \times e_3) = -u^2$$

$$\widetilde{B}_{11} = \varphi(\tilde{b}_1, b_1) = u \cdot (e_2 \times e_2) = 0$$

$$\widetilde{B}_{22} = \varphi(\tilde{b}_2, b_2) = u \cdot ((e_1 + e_3) \times (e_1 + e_3)) = 0$$

$$\widetilde{B}_{33} = \varphi(\tilde{b}_3, b_3) = u \cdot (e_3 \times e_3) = 0$$

□

Example 3.33 If, in the definition of the scalar triple product, instead of fixing a vector $a \in \mathbb{R}$, we let the vector vary, we have a function $\varphi : \mathbb{R}^3 \times \mathbb{R}^3 \times \mathbb{R}^3 \to \mathbb{R}$, defined by

$$\varphi(u, v, w) := u \cdot (v \times w) = \det \begin{bmatrix} u \\ v \\ w \end{bmatrix}.$$

One can verify that such function is trilinear, that is linear in each of the three variables separately.

The components T_{ijk} of this trilinear form are simply given by the sign of the corresponding permutation:

$$\varphi = \text{sign}(i, j, k)\beta^i \otimes \beta^j \otimes \beta^k = \beta^1 \otimes \beta^2 \otimes \beta^3 - \beta^1 \otimes \beta^3 \otimes \beta^2 + \beta^3 \otimes \beta^1 \otimes \beta^2$$
$$- \beta^3 \otimes \beta^2 \otimes \beta^1 + \beta^2 \otimes \beta^3 \otimes \beta^1 - \beta^2 \otimes \beta^1 \otimes \beta^3,$$

where the sign of the permutation is given by

$$\text{sign}(i, j, k) := \begin{cases} +1 & \text{if } (i, j, k) = (1, 2, 3), (2, 3, 1) \text{ or } (3, 1, 2) \\ & \quad (\textbf{even permutations of}(1, 2, 3)) \\ -1 & \text{if } (i, j, k) = (1, 3, 2), (2, 1, 3) \text{ or } (3, 2, 1) \\ & \quad (\textbf{odd permutations of}(1, 2, 3)) \\ 0 & \text{otherwise.} \end{cases}$$

□

3.3 Multilinear Forms

Example 3.34 In general, the **determinant** defines an n-linear form in $\mathbb{R}^n$ by

$$\varphi : \underbrace{\mathbb{R}^n \times \ldots \times \mathbb{R}^n}_{n \text{ factors}} \longrightarrow \mathbb{R}, \qquad \varphi(v_1, \ldots, v_n) := \det \begin{bmatrix} v_1 \\ \vdots \\ v_n \end{bmatrix},$$

where we compute the determinant of the square matrix with rows (equivalently, columns) given by the n vectors. Multilinearity is a fundamental property of the determinant.

In this case, the components of this multilinear form are also given by the permutation signs:

$$\varphi = \text{sign}(i_1, \ldots, i_n) \beta^{i_1} \otimes \ldots \otimes \beta^{i_n},$$

where

$$\text{sign}(i_1, \ldots, i_n) := \begin{cases} +1 & \text{if } (i_1, \ldots, i_n) \text{ is an even permutation of } (1, \ldots, n) \\ -1 & \text{if } (i_1, \ldots, i_n) \text{ is an odd permutation of } (1, \ldots, n) \\ 0 & \text{otherwise.} \end{cases}$$

A permutation of $(1, 2, \ldots, n)$ is called an **even permutation**, if it is obtained from $(1, 2, \ldots, n)$ by an even number of two-element swaps; otherwise it is called an **odd permutation**. □

3.3.3 Tensor Product of Multilinear Forms

Let

$$T : \underbrace{V \times \cdots \times V}_{k \text{ times}} \to \mathbb{R} \qquad \text{and} \qquad U : \underbrace{V \times \cdots \times V}_{\ell \text{ times}} \to \mathbb{R}$$

be, respectively, a k-linear and an ℓ-linear form. Then the **tensor product** of T and U is the function

$$T \otimes U : \underbrace{V \times \cdots \times V}_{k+\ell \text{ times}} \to \mathbb{R}$$

defined by

$$T \otimes U(v_1, \ldots, v_{k+\ell}) := T(v_1, \ldots, v_k) U(v_{k+1}, \ldots, v_{k+\ell}).$$

This is a $(k + \ell)$-linear form. Equivalently, this is saying that the tensor product of a tensor of type $(0, k)$ and a tensor of type $(0, \ell)$ is a tensor of type $(0, k + \ell)$. Later we will see how this product extends to more general tensors.

Inner Products   4

This chapter introduces inner products as a special case of bilinear forms. It discusses the representation of an inner product via a symmetric positive definite matrix, the notion of reciprocal basis, and how the presence of an inner product blurs covariance and contravariance.

4.1 Definitions and First Properties

Inner products add an important structure to a vector space, as for example they allow to compute the length of a vector and they provide a canonical identification between the vector space V and its dual V^*.

4.1.1 Inner Products and Their Related Notions

DEFINITION 4.1. An **inner product** $g : V \times V \to \mathbb{R}$ on a real vector space V is a *bilinear form* on V that is

(1) *symmetric*, that is $g(v, w) = g(w, v)$ for all $v, w \in V$ and
(2) *positive definite*, that is $g(v, v) \geq 0$ for all $v \in V$, and $g(v) = 0$ if and only if $v = 0$.

EXERCISE 4.2. Let $V = \mathbb{R}^3$. Verify that the dot product $\varphi(v, w) := v \cdot w$, defined as

$$v \cdot w = v^i w^j \delta_{ij},$$

where $v = \begin{bmatrix} v^1 \\ v^2 \\ v^3 \end{bmatrix}$ and $w = \begin{bmatrix} w^1 \\ w^2 \\ w^3 \end{bmatrix}$ is an inner product. This is called the **standard inner product**.

EXERCISE 4.3. Determine whether the following bilinear forms $\varphi : \mathbb{R}^n \times \mathbb{R}^n \to \mathbb{R}$ are inner products, by verifying whether they are symmetric and positive definite (the formulas are throughout defined for all $v, w \in \mathbb{R}^n$):

(1) $\varphi(v, w) := -v \cdot w$;
(2) $\varphi(v, w) := v \cdot w + 2v^1 w^2$;
(3) $\varphi(v, w) := v^1 w^1$;
(4) $\varphi(v, w) := v \cdot w - 2v^1 w^1$;
(5) $\varphi(v, w) := v \cdot w + 2v^1 w^1$;
(6) $\varphi(v, w) := v \cdot 3w$.

EXERCISE 4.4. Let $V := P_2(\mathbb{R})$ be the vector space of polynomials of degree ≤ 2. Determine whether the following bilinear forms are inner products, by verifying whether they are symmetric and positive definite:

(1) $\varphi(p, q) := \int_0^1 p(x) q(x) dx$;
(2) $\varphi(p, q) := \int_0^1 p'(x) q'(x) dx$;
(3) $\varphi(p, q) := \int_3^\pi e^x p(x) q(x) dx$;
(4) $\varphi(p, q) := p(1)q(1) + p(2)q(2)$;
(5) $\varphi(p, q) := p(1)q(1) + p(2)q(2) + p(3)q(3)$.
(6) $\varphi(p, q) := p(1)q(2) + p(2)q(3) + p(3)q(1)$.

DEFINITION 4.5. Let $g : V \times V \to \mathbb{R}$ be an inner product on V.

(1) The **norm** (or **magnitude** or **length**), $\|v\|$ of a vector $v \in V$ is defined as
$$\|v\| := \sqrt{g(v, v)}.$$
(2) A vector $v \in V$ is **unit vector** if $\|v\| = 1$;
(3) Two vectors $v, w \in V$ are **orthogonal** (that is, **perpendicular** denoted $v \perp w$), if $g(v, w) = 0$;
(4) Two vectors $v, w \in V$ are **orthonormal** if they are orthogonal and $\|v\| = \|w\| = 1$;

4.1 Definitions and First Properties

(5) A basis $\mathcal{B}$ of V is an **orthonormal basis** if $b_1, \ldots, b_n$ are pairwise orthonormal vectors, that is

$$g(b_i, b_j) = \delta_{ij} := \begin{cases} 1 & \text{if } i = j \\ 0 & \text{if } i \neq j, \end{cases} \qquad (4.1)$$

for all $i, j = 1 \ldots, n$. The condition for $i = j$ implies that an orthonormal basis consists of unit vectors, while the one for $i \neq j$ implies that it consists of pairwise orthogonal vectors.

Example 4.6 (1) Let $V = \mathbb{R}^n$ and g the standard inner product. The standard basis $\mathcal{B} = \{e_1, \ldots, e_n\}$ is an orthonormal basis with respect to the standard inner product.
(2) Let $V = P_2(\mathbb{R})$ and let $g(p, q) := \int_{-1}^{1} p(x)q(x)dx$. Check that the basis

$$\mathcal{B} = \{p_1, p_2, p_3\},$$

where

$$p_1(x) := \tfrac{1}{\sqrt{2}}, \quad p_2(x) := \sqrt{\tfrac{3}{2}} x, \quad p_3(x) := \sqrt{\tfrac{5}{8}}(3x^2 - 1),$$

is an orthonormal basis with respect to the inner product g. Up to scaling, p_1, p_2, p_3 are the first three **Legendre polynomials**. $\square$

An inner product g on a vector space V induces a **metric**[1] on V, where the distance between vectors $v, w \in V$ is given by

$$d(v, w) := \|v - w\|.$$

4.1.2 Symmetric Matrices and Quadratic Forms

Recall that a matrix $S \in M_{n \times n}(\mathbb{R})$ is **symmetric** if $S = {}^t S$, that is if

$$S = \begin{bmatrix} * & a & b & \cdot^{\cdot^{\cdot}} \\ a & * & c & \cdot^{\cdot^{\cdot}} \\ b & c & * & \cdot^{\cdot^{\cdot}} \\ \cdot_{\cdot_{\cdot}} & \cdot_{\cdot_{\cdot}} & & * \end{bmatrix}.$$

---

[1] An inner product induces a *norm* and a norm induces a *metric* on a vector space. However, the converses do not hold.

Moreover, if S is symmetric, then

(1) S is **positive definite** if ${}^t v S v > 0$ for all $v \in \mathbb{R}^n \setminus \{0\}$;
(2) S is **negative definite** if ${}^t v S v < 0$ for all $v \in \mathbb{R}^n \setminus \{0\}$;
(3) S is **positive semidefinite** if ${}^t v S v \geq 0$ for all $v \in \mathbb{R}^n$;
(4) S is **negative semidefinite** if ${}^t v S v \leq 0$ for all $v \in \mathbb{R}^n$;
(5) S is **indefinite** if ${}^t v S v$ takes both positive and negative values for different $v \in \mathbb{R}^n$.

DEFINITION 4.7. A **quadratic form** $Q : \mathbb{R}^n \to \mathbb{R}$ is a homogeneous quadratic polynomial in n variables:

$$Q(x^1, \ldots, x^n) = Q_{ij} x^i x^j, \qquad \text{where } Q_{ij} \in \mathbb{R}.$$

To any symmetric matrix S corresponds a **quadratic form** $Q_S : \mathbb{R}^n \to \mathbb{R}$ defined by

$$Q_S(v) = \underbrace{{}^t v S v = \begin{bmatrix} v^1 & \ldots & v^n \end{bmatrix} S \begin{bmatrix} v^1 \\ \vdots \\ v^n \end{bmatrix}}_{\text{matrix notation}} = \underbrace{v^i v^j S_{ij}}_{\text{Einstein notation}} . \qquad (4.2)$$

Note that Q is *not* linear in v.

Let S be a symmetric matrix and Q_S be the corresponding quadratic form. The notion of positive definiteness, etc. for S translates into corresponding properties for Q_S, namely:

(1) Q is **positive definite** if $Q(v) > 0$ for all $v \in \mathbb{R}^n \setminus \{0\}$;
(2) Q is **negative definite** if $Q(v) < 0$ for all $v \in \mathbb{R}^n \setminus \{0\}$;
(3) Q is **positive semidefinite** if $Q(v) \geq 0$ for all $v \in \mathbb{R}^n$;
(4) Q is **negative semidefinite** if $Q(v) \leq 0$ for all $v \in \mathbb{R}^n$;
(5) Q is **indefinite** if $Q(v)$ takes both positive and negative values.

Example 4.8 We consider $\mathbb{R}^2$ with the standard basis $\mathcal{E}$ and the quadratic form[2] $Q(v) := v^1 v^1 - v^2 v^2$, where $v = \begin{bmatrix} v^1 \\ v^2 \end{bmatrix}$. The symmetric matrix corresponding

---

[2] Here, we avoid the usual notation for squares, because of the possible confusion with upper indices.

4.1 Definitions and First Properties

to Q is $S := \begin{bmatrix} 1 & 0 \\ 0 & -1 \end{bmatrix}$. If $v = \begin{bmatrix} v^1 \\ 0 \end{bmatrix}$, then $Q(v) = 1 > 0$, if $v = \begin{bmatrix} 0 \\ v^2 \end{bmatrix}$, then $Q(v) = -1 < 0$, but any vector for which $v^1 = v^2$ has the property that $Q(v) = 0$. □

To find out the type of a symmetric matrix S (or, equivalently of a quadratic form Q_S) it is enough to look at the eigenvalues of S, namely:

(1) S and Q_S are *positive definite* when all eigenvalues of S are positive;
(2) S and Q_S are *negative definite* when all eigenvalues of S are negative;
(3) S and Q_S are *positive semidefinite* when all eigenvalues of S are non-negative;
(4) S and Q_S are *negative semidefinite* when all eigenvalues of S are non-positive;
(5) S and Q_S are *indefinite* when S has both positive and negative eigenvalues.

The reason this makes sense is the same reason for which we need to restrict our attention to symmetric matrices and lies in the celebrated Spectral Theorem from Linear Algebra (see, for instance, [13, §5.5]):

Theorem 4.9 (Spectral Theorem) *Any $n \times n$ symmetric matrix S has the following properties:*

(a) it has only real eigenvalues;
(b) it is diagonalizable;
(c) it admits an orthonormal eigenbasis, that is, a basis $\{b_1, \ldots, b_n\}$ of $\mathbb{R}^n$ such that the b_j are orthonormal and are eigenvectors of S.

4.1.3 Inner Products vs. Symmetric Positive Definite Matrices

Let $\mathcal{B} = \{b_1, \ldots, b_n\}$ be a basis of V and g an inner product. The **components of g with respect to $\mathcal{B}$** are

$$g_{ij} := g(b_i, b_j). \tag{4.3}$$

Let G be the matrix with entries g_{ij}

$$G = \begin{bmatrix} g_{11} & \cdots & g_{1n} \\ \vdots & \ddots & \vdots \\ g_{n1} & \cdots & g_{nn} \end{bmatrix}. \tag{4.4}$$

We claim that G is symmetric and positive definite. In fact:

(1) Since g is *symmetric*, then for $1 \le i, j \le n$,

$$g_{ij} = g(b_i, b_j) = g(b_j, b_i) = g_{ji} \quad \Longrightarrow \quad G \text{ is a } \textit{symmetric} \text{ matrix};$$
(4.5)

(2) Since g is *positive definite*, then G is *positive definite* as a symmetric matrix. In fact, let $v = v^i b_i$, $w = w^j b_j \in V$ be two vectors. Then, using the bilinearity of g, the definition in Eq. (4.3) and Einstein notation, we have:

$$g(v, w) = g(v^i b_i, w^j b_j) = v^i w^j \underbrace{g(b_i, b_j)}_{g_{ij}} = v^i w^j g_{ij}$$

or, in matrix notation,

$$g(v, w) = {}^t[v]_\mathcal{B} G [w]_\mathcal{B} = \begin{bmatrix} v^1 & \ldots & v^n \end{bmatrix} G \begin{bmatrix} w^1 \\ \vdots \\ w^n \end{bmatrix}.$$

Conversely, if S is a symmetric positive definite matrix and $\mathcal{B} = \{b_1, \ldots, b_n\}$ is a basis of V, then the assignment

$$V \times V \longrightarrow \mathbb{R}, \qquad (v, w) \longmapsto {}^t[v]_\mathcal{B} S [w]_\mathcal{B}$$

defines a map that is seen to be bilinear, symmetric and positive definite, hence an inner product.

4.1.4 Orthonormal Bases

Suppose that there is a basis $\mathcal{B} = \{b_1, \ldots, b_n\}$ of V consisting of orthonormal vectors with respect to an inner product g, so that

$$g_{ij} = \delta_{ij};$$

cf. Definition 4.5(5) and Eq. (4.3). In other words, the symmetric matrix corresponding to the inner product g in the basis consisting of orthonormal vectors is the identity matrix. Moreover, we have

$$g(v, w) = v^i w^j g_{ij} = v^i w^j \delta_{ij} = v^1 w^1 + \cdots + v^n w^n,$$

4.1 Definitions and First Properties

so that, in the case $v = w$, we get

$$\|v\|^2 = g(v, v) = v^i v^j \delta_{ij} = v^1 v^1 + \cdots + v^n v^n.$$

We thus deduce that:

Fact 4.10 Any inner product g can be expressed in the standard form

$$g(v, w) = v^i w^j \delta_{ij} = v^1 w^1 + \cdots + v^n w^n,$$

as long as $[v]_B = \begin{pmatrix} v^1 \\ \vdots \\ v^n \end{pmatrix}$ and $[w]_B = \begin{pmatrix} w^1 \\ \vdots \\ w^n \end{pmatrix}$ are the coordinates of v and w with respect to an orthonormal basis B for g.

Example 4.11 Let g be an inner product of $\mathbb{R}^3$ with respect to which

$$\widetilde{B} := \{ \underbrace{\begin{bmatrix} 1 \\ 0 \\ 0 \end{bmatrix}}_{\tilde{b}_1}, \underbrace{\begin{bmatrix} 1 \\ 1 \\ 0 \end{bmatrix}}_{\tilde{b}_2}, \underbrace{\begin{bmatrix} 1 \\ 1 \\ 1 \end{bmatrix}}_{\tilde{b}_3} \}$$

is an orthonormal basis. We want to express g with respect to the standard basis of $\mathbb{R}^3$,

$$\mathcal{E} := \{ \underbrace{\begin{bmatrix} 1 \\ 0 \\ 0 \end{bmatrix}}_{e_1}, \underbrace{\begin{bmatrix} 0 \\ 1 \\ 0 \end{bmatrix}}_{e_2}, \underbrace{\begin{bmatrix} 0 \\ 0 \\ 1 \end{bmatrix}}_{e_3} \}.$$

The matrices of the change of basis are

$$L := L_{\mathcal{E}\widetilde{B}} = \begin{bmatrix} 1 & 1 & 1 \\ 0 & 1 & 1 \\ 0 & 0 & 1 \end{bmatrix} \quad \text{and} \quad \Lambda = L^{-1} = \begin{bmatrix} 1 & -1 & 0 \\ 0 & 1 & -1 \\ 0 & 0 & 1 \end{bmatrix}.$$

Since g is a bilinear form, we saw in Eq. (3.15) that its matrices with respect to the bases B and $\mathcal{E}$ are related by the formula

$$\widetilde{G} = {}^t L G L.$$

Since the basis $\widetilde{\mathcal{B}}$ is orthonormal with respect to g, the associated matrix $\widetilde{G}$ is the identity matrix, so that

$$G = {}^t\Lambda \widetilde{G} \Lambda = {}^t\Lambda \Lambda$$
$$= \begin{bmatrix} 1 & 0 & 0 \\ -1 & 1 & 0 \\ 0 & -1 & 1 \end{bmatrix} \begin{bmatrix} 1 & -1 & 0 \\ 0 & 1 & -1 \\ 0 & 0 & 1 \end{bmatrix} = \begin{bmatrix} 1 & -1 & 0 \\ -1 & 2 & -1 \\ 0 & -1 & 2 \end{bmatrix}. \quad (4.6)$$

It follows that, with respect to the standard basis, g is given by

$$g(v, w) = \begin{pmatrix} v^1 & v^2 & v^3 \end{pmatrix} \begin{bmatrix} 1 & -1 & 0 \\ -1 & 2 & -1 \\ 0 & -1 & 2 \end{bmatrix} \begin{pmatrix} w^1 \\ w^2 \\ w^3 \end{pmatrix} \quad (4.7)$$
$$= v^1 w^1 - v^1 w^2 - v^2 w^1 + 2v^2 w^2 - v^2 w^3 - v^3 w^2 + 2v^3 w^3.$$

$\square$

EXERCISE 4.12. Verify the formula (4.7) for the inner product g w.r.t. the basis $\widetilde{\mathcal{B}}$ by applying the change of basis matrix directly to the coordinate vectors $[v]_\mathcal{E}$, $[w]_\mathcal{E}$.

Remark 4.13 Norms and the value of the inner product of vectors depend *only* on the choice of g, *not* on the choice of basis; different coordinate expressions yield the same result:

$$g(v, w) = {}^t[v]_\mathcal{B} G [v]_\mathcal{B} = {}^t[v]_{\widetilde{\mathcal{B}}} \widetilde{G} [v]_{\widetilde{\mathcal{B}}}.$$

Example 4.14 We verify the assertion of the previous remark with the inner product in Example 4.11. Let $v, w \in \mathbb{R}^3$ such that

$$[v]_\mathcal{E} = \begin{pmatrix} v^1 \\ v^2 \\ v^3 \end{pmatrix} = \begin{pmatrix} 3 \\ 2 \\ 1 \end{pmatrix} \quad \text{and} \quad [v]_{\widetilde{\mathcal{B}}} = \begin{pmatrix} \tilde{v}^1 \\ \tilde{v}^2 \\ \tilde{v}^3 \end{pmatrix} = L^{-1} \begin{pmatrix} v^1 \\ v^2 \\ v^3 \end{pmatrix} = \begin{pmatrix} 1 \\ 1 \\ 1 \end{pmatrix}$$

and

$$[w]_\mathcal{E} = \begin{pmatrix} w^1 \\ w^2 \\ w^3 \end{pmatrix} = \begin{pmatrix} 1 \\ 2 \\ 3 \end{pmatrix} \quad \text{and} \quad [w]_{\widetilde{\mathcal{B}}} = \begin{pmatrix} \tilde{w}^1 \\ \tilde{w}^2 \\ \tilde{w}^3 \end{pmatrix} = L^{-1} \begin{pmatrix} w^1 \\ w^2 \\ w^3 \end{pmatrix} = \begin{pmatrix} -1 \\ -1 \\ 3 \end{pmatrix}.$$

4.1 Definitions and First Properties

Then with respect to the basis $\widetilde{\mathcal{B}}$ we have that

$$g(v, w) = 1 \cdot (-1) + 1 \cdot (-1) + 1 \cdot 3 = 1,$$

and also with respect to the basis $\mathcal{E}$

$$g(v, w) = 3 \cdot 1 - 3 \cdot 2 - 2 \cdot 1 + 2 \cdot 2 \cdot 2 - 2 \cdot 3 - 1 \cdot 2 + 2 \cdot 1 \cdot 3 = 1.$$

□

EXERCISE 4.15. Verify that $\|v\| = \sqrt{3}$ and $\|w\| = \sqrt{11}$, when computed with respect to either basis.

Let $\mathcal{B} = \{b_1, \ldots, b_n\}$ be an orthonormal basis and let $v = v^i b_i$ be a vector in V.

▶ Then the coordinates v^i of the vector v with respect to this *orthonormal* basis can be obtained by computing the inner product of v with the basis vectors:

$$g(v, b_j) = g(v^i b_i, b_j) = v^i g(b_i, b_j) = v^i \delta_{ij} = v^j.$$

This is particularly nice, so that we have to make sure that we remember how to construct an orthonormal basis from a given arbitrary basis.

Recall The **Gram-Schmidt orthogonalization process** is a recursive process that allows us to obtain an orthonormal basis starting from an arbitrary one. Let $\mathcal{B} = \{b_1, \ldots, b_n\}$ be an arbitrary basis, let $g : V \times V \to \mathbb{R}$ be an inner product and $\|\cdot\|$ the corresponding norm.

We start by recalling that the **orthogonal projection**, denoted $\mathrm{proj}_{b_k} v$, of a vector $v \in V$ onto the line spanned by a non-zero vector b_k is defined as

$$\mathrm{proj}_{b_k} v := \frac{g(v, b_k)}{g(b_k, b_k)} b_k. \tag{4.8}$$

The vector $\mathrm{proj}_{b_k} v$ is clearly parallel (i.e., proportional) to b_k and the following exercise shows that the complement $v - \mathrm{proj}_{b_k} v$ is indeed orthogonal to b_k (Fig. 4.1).

Fig. 4.1 Orthogonal projection of vector v onto line spanned by unit vector b_k

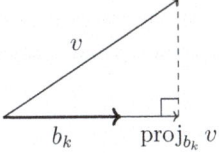

Therefore, with respect to the line spanned by b_k, we have the decomposition

$$v = \underbrace{\mathrm{proj}_{b_k} v}_{\text{parallel}} + \underbrace{v - \mathrm{proj}_{b_k} v}_{\text{orthogonal}} .$$

EXERCISE 4.16. With $\mathrm{proj}_{b_k} v$ defined as in (4.8), check that we have

$$(v - \mathrm{proj}_{b_k} v) \perp b_k ,$$

where the orthogonality is meant with respect to the inner product g.

Given the basis $\mathcal{B} = \{b_1, \ldots, b_n\}$ of V, we will find an orthonormal basis. We start by defining

$$u_1 := \frac{1}{\|b_1\|} b_1 .$$

Next, observe that $g(b_2, u_1) u_1$ is the projection of the vector b_2 in the direction of u_1. It follows that

$$b_2^\perp := b_2 - g(b_2, u_1) u_1$$

is a vector orthogonal to u_1, but not necessarily of unit norm. Hence we set

$$u_2 := \frac{1}{\|b_2^\perp\|} b_2^\perp .$$

Likewise, $g(b_3, u_1) u_1 + g(b_3, u_2) u_2$ is the projection of b_3 on the plane generated by u_1 and u_2, so that the difference

$$b_3^\perp := b_3 - g(b_3, u_1) u_1 - g(b_3, u_2) u_2$$

is orthogonal both to u_1 and to u_2. Set

$$u_3 := \frac{1}{\|b_3^\perp\|} b_3^\perp .$$

Continuing this way until we have exhausted all elements of the basis $\mathcal{B}$, we obtain an orthonormal basis $\{u_1, \ldots, u_n\}$ (Fig. 4.2).

4.1 Definitions and First Properties

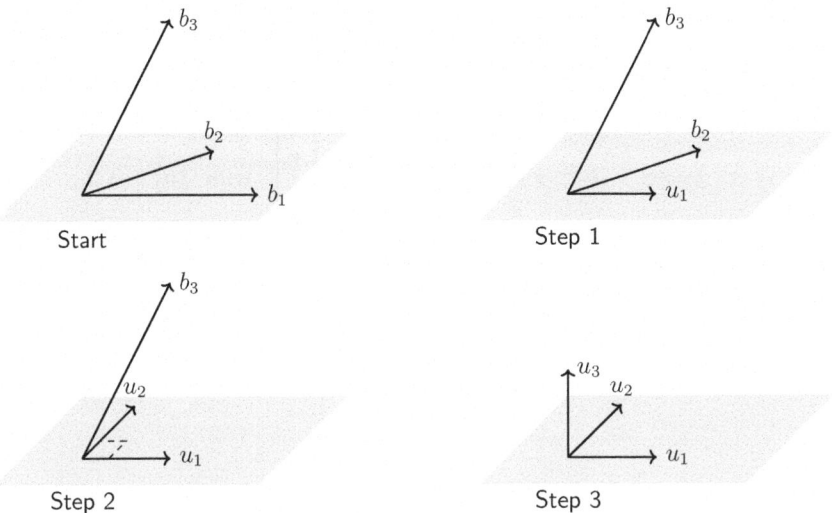

Fig. 4.2 Gram-Schmidt process on three vectors

Example 4.17 Let V be the subspace of $\mathbb{R}^4$ spanned by

$$b_1 = \begin{bmatrix} 1 \\ 1 \\ -1 \\ -1 \end{bmatrix} \quad b_2 = \begin{bmatrix} 2 \\ 2 \\ 0 \\ 0 \end{bmatrix} \quad b_3 = \begin{bmatrix} 1 \\ 1 \\ 1 \\ 0 \end{bmatrix}.$$

(One can check that b_1, b_2, b_3 are linearly independent and hence form a basis of V.) We look for an orthonormal basis of V with respect to the standard inner product $\langle \cdot, \cdot \rangle$. Since

$$\|b_1\| = \left(1^2 + 1^2 + (-1)^2 + (-1)^2\right)^{1/2} = 2,$$

we find

$$u_1 := \tfrac{1}{2} b_1 = \tfrac{1}{2} \begin{bmatrix} 1 \\ 1 \\ -1 \\ -1 \end{bmatrix}.$$

Moreover,

$$\langle b_2, u_1 \rangle = \tfrac{1}{2}(1+1) = 2$$

$$\implies \quad b_2^\perp := b_2 - \langle b_2, u_1 \rangle u_1 = \begin{bmatrix} 1 \\ 1 \\ 1 \\ 1 \end{bmatrix} \quad \text{with } \|b_2^\perp\| = 2,$$

so that we find

$$u_2 := \tfrac{1}{2} b_2^\perp = \tfrac{1}{2} \begin{bmatrix} 1 \\ 1 \\ 1 \\ 1 \end{bmatrix}.$$

Finally,

$$\langle b_3, u_1 \rangle = \tfrac{1}{2}(1+1-1) = \tfrac{1}{2} \quad \text{and} \quad \langle b_3, u_2 \rangle = \tfrac{1}{2}(1+1+1) = \tfrac{3}{2}$$

imply that

$$b_3^\perp := b_3 - \langle b_3, u_1 \rangle u_1 - \langle b_3, u_2 \rangle u_2 = \begin{bmatrix} 0 \\ 0 \\ \tfrac{1}{2} \\ -\tfrac{1}{2} \end{bmatrix} \quad \text{with } \|b_3^\perp\| = \tfrac{\sqrt{2}}{2},$$

so that we have

$$u_3 := \tfrac{\sqrt{2}}{2} b_3^\perp = \tfrac{\sqrt{2}}{2} \begin{bmatrix} 0 \\ 0 \\ 1 \\ -1 \end{bmatrix}.$$

□

4.2 Reciprocal Basis

4.2.1 Definition and Examples

Let $g : V \times V \to \mathbb{R}$ be an inner product and $\mathcal{B} = \{b_1, \ldots, b_n\}$ any basis of V.

4.2 Reciprocal Basis

DEFINITION 4.18. The **reciprocal basis** of V with respect to the inner product g and the given $\mathcal{B}$ is the basis of V, denoted by

$$\mathcal{B}^g = \{b^1, \ldots, b^n\}$$

that satisfies

$$\boxed{g(b^i, b_j) = \delta^i_j}. \qquad (4.9)$$

Note that, while it is certainly possible to define a set of $n = \dim V$ vectors satisfying Eq. (4.9), we need to justify the fact that we call it a *basis*. This will be done in Claim 4.22.

Remark 4.19 In general, $\mathcal{B}^g \neq \mathcal{B}$ and in fact, because of Definition 4.5(5),

$$\mathcal{B} = \mathcal{B}^g \quad \Longleftrightarrow \quad \mathcal{B} \text{ is an orthonormal basis.}$$

Example 4.20 Let g be the inner product defined in (4.7) in Example 4.11 and let $\mathcal{E}$ the standard basis of $\mathbb{R}^3$. We want to find the reciprocal basis $\mathcal{E}^g$, that is we want to find $\mathcal{E}^g := \{e^1, e^2, e^3\}$ such that

$$g(e^i, e_j) = \delta^i_j.$$

If G is the matrix of the inner product given in (4.6), using the matrix notation and considering e^j as a row vector and e_i as a column vector for $i, j = 1, 2, 3$, we have

$$\begin{bmatrix} - & {}^t e^i & - \end{bmatrix} G \begin{bmatrix} | \\ e_j \\ | \end{bmatrix} = \delta^i_j.$$

Letting i and j vary from 1 to 3, we obtain

$$\begin{bmatrix} - & {}^t e^1 & - \\ - & {}^t e^2 & - \\ - & {}^t e^3 & - \end{bmatrix} \begin{bmatrix} 1 & -1 & 0 \\ -1 & 2 & -1 \\ 0 & -1 & 2 \end{bmatrix} \begin{bmatrix} | & | & | \\ e_1 & e_2 & e_3 \\ | & | & | \end{bmatrix} = \begin{bmatrix} 1 & 0 & 0 \\ 0 & 1 & 0 \\ 0 & 0 & 1 \end{bmatrix},$$

from which we conclude that

$$\begin{bmatrix} - & {}^t e^1 & - \\ - & {}^t e^2 & - \\ - & {}^t e^3 & - \end{bmatrix} = \begin{bmatrix} | & | & | \\ e_1 & e_2 & e_3 \\ | & | & | \end{bmatrix}^{-1} \begin{bmatrix} 1 & -1 & 0 \\ -1 & 2 & -1 \\ 0 & -1 & 2 \end{bmatrix}^{-1}$$

$$= \begin{bmatrix} | & | & | \\ e_1 & e_2 & e_3 \\ | & | & | \end{bmatrix} \begin{bmatrix} 3 & 2 & 1 \\ 2 & 2 & 1 \\ 1 & 1 & 1 \end{bmatrix} = \begin{bmatrix} 3 & 2 & 1 \\ 2 & 2 & 1 \\ 1 & 1 & 1 \end{bmatrix}.$$

Therefore,

$$e^1 = \begin{bmatrix} 3 \\ 2 \\ 1 \end{bmatrix}, \quad e^2 = \begin{bmatrix} 2 \\ 2 \\ 1 \end{bmatrix}, \quad e^3 = \begin{bmatrix} 1 \\ 1 \\ 1 \end{bmatrix}. \tag{4.10}$$

Observe that in order to compute G^{-1} we used the Gauss–Jordan elimination method

$$\begin{bmatrix} 1 & -1 & 0 & | & 1 & 0 & 0 \\ -1 & 2 & -1 & | & 0 & 1 & 0 \\ 0 & -1 & 2 & | & 0 & 0 & 1 \end{bmatrix} \rightsquigarrow \begin{bmatrix} 1 & -1 & 0 & | & 1 & 0 & 0 \\ 0 & 1 & -1 & | & 1 & 1 & 0 \\ 0 & -1 & 2 & | & 0 & 0 & 1 \end{bmatrix}$$

$$\rightsquigarrow \begin{bmatrix} 1 & 0 & -1 & | & 2 & 1 & 0 \\ 0 & 1 & -1 & | & 1 & 1 & 0 \\ 0 & 0 & 1 & | & 1 & 1 & 1 \end{bmatrix} \rightsquigarrow \begin{bmatrix} 1 & 0 & 0 & | & 3 & 2 & 1 \\ 0 & 1 & 0 & | & 2 & 2 & 1 \\ 0 & 0 & 1 & | & 1 & 1 & 1 \end{bmatrix}.$$

□

EXERCISE 4.21. Again in the situation of Examples 4.11 and 4.20, let g be an inner product on $\mathbb{R}^3$, let $\mathcal{E} = \{e_1, e_2, e_3\}$ be the standard basis and let

$$\widetilde{\mathcal{B}} := \{ \underbrace{\begin{bmatrix} 1 \\ 0 \\ 0 \end{bmatrix}}_{\tilde{b}_1}, \underbrace{\begin{bmatrix} 1 \\ 1 \\ 0 \end{bmatrix}}_{\tilde{b}_2}, \underbrace{\begin{bmatrix} 1 \\ 1 \\ 1 \end{bmatrix}}_{\tilde{b}_3} \}$$

be an orthonormal basis with respect to g.

(1) Compute $[\tilde{b}_1]_{\widetilde{\mathcal{B}}}$, $[\tilde{b}_2]_{\widetilde{\mathcal{B}}}$ and $[\tilde{b}_3]_{\widetilde{\mathcal{B}}}$.
(2) Compute the matrix $G_{\widetilde{\mathcal{B}}}$ associated to g with respect to the basis $\widetilde{\mathcal{B}}$, and the matrix $G_{\mathcal{E}}$ associated to g with respect to the basis $\mathcal{E}$.
(3) We denote by $\mathcal{E}^g = \{e^1, e^2, e^3\}$ and $\widetilde{\mathcal{B}}^g = \{\tilde{b}^1, \tilde{b}^2, \tilde{b}^3\}$ the reciprocal bases respectively of $\mathcal{E}$ and $\widetilde{\mathcal{B}}$. By Remark 4.13,

$$g(\tilde{b}^i, \tilde{b}_j) = \delta^i_j \quad \text{and} \quad g(e^i, e_j) = \delta^i_j$$

are independent of the choice of the basis. It follows that:
(a) $\delta^i_j = g(\tilde{b}^i, \tilde{b}_j) = [{}^t\tilde{b}^i]_{\widetilde{\mathcal{B}}} G_{\widetilde{\mathcal{B}}} [\tilde{b}_j]_{\widetilde{\mathcal{B}}}$;
(b) $\delta^i_j = g(\tilde{b}^i, \tilde{b}_j) = [{}^t\tilde{b}^i]_{\mathcal{E}} G_{\mathcal{E}} [\tilde{b}_j]_{\mathcal{E}}$;
(c) $\delta^i_j = g(e^i, e_j) = [{}^t e^i]_{\widetilde{\mathcal{B}}} G_{\widetilde{\mathcal{B}}} [e_j]_{\widetilde{\mathcal{B}}}$;
(d) $\delta^i_j = g(e^i, e_j) = [{}^t e^i]_{\mathcal{E}} G_{\mathcal{E}} [e_j]_{\mathcal{E}}$.

4.2 Reciprocal Basis

Using (1), (2) and the appropriate formula among (a), (b), (c) and (d), compute $[\tilde{b}^i]_{\tilde{\mathcal{B}}}$, $[\tilde{b}^i]_{\mathcal{E}}$, $[e^i]_{\mathcal{E}}$ and $[e^i]_{\tilde{\mathcal{B}}}$. For some of these, one will probably want to apply the same technique as in Example 4.20.

4.2.2 Properties of Reciprocal Bases

CLAIM 4.22 Given a vector space V with a basis $\mathcal{B}$ and an inner product $g : V \times V \to \mathbb{R}$, a reciprocal basis *exists* and is *unique*.

As we pointed out right after the definition of reciprocal basis, what this claim really says is that there is a set of vectors $\{b^1, \ldots, b^n\}$ in V that satisfy Eq. (4.9), that form a basis and that this basis is unique.

Proof Let $\mathcal{B} = \{b_1, \ldots, b_n\}$ be the given basis. Any other basis $\{b^1, \ldots, b^n\}$ is related to $\mathcal{B}$ by the relation

$$b^i = M^{ij} b_j \tag{4.11}$$

for some *invertible* matrix M. We want to show that there exists a *unique* matrix M such that, when Eq. (4.11) is plugged into $g(b^i, b_j)$, we have

$$g(b^i, b_j) = \delta^i_j . \tag{4.12}$$

From Eqs. (4.11) and (4.12) we obtain

$$\delta^i_j = g(b^i, b_j) = g(M^{ik} b_k, b_j) = M^{ik} g(b_k, b_j) = M^{ik} g_{kj} ,$$

which, in matrix notation, becomes

$$I = MG ,$$

where G is the matrix of g with respect to $\mathcal{B}$ whose entries are g_{ij} as in (4.4). Since G is invertible because it is positive definite, then $M = G^{-1}$ exists and is unique. $\square$

Remark 4.23 Note that in the course of the proof we have found that, since $M = L_{\mathcal{B}\mathcal{B}^g}$, then

$$\boxed{G = (L_{\mathcal{B}\mathcal{B}^g})^{-1} = L_{\mathcal{B}^g \mathcal{B}}} .$$

We denote with g^{ij} the entries of $M = G^{-1}$. From the above discussion, it follows that with this notation

$$\boxed{g^{ik}g_{kj} = \delta^i_j} \tag{4.13}$$

as well as

$$\boxed{b^i = g^{ij}b_j}, \tag{4.14}$$

or[3]

$$\boxed{(b^1 \ldots b^n) = (b_1 \ldots b_n)\, G^{-1}}. \tag{4.15}$$

(note that G^{-1} has to be multiplied on the right). This is consistent with the findings in Sect. 2.3.2. We can now compute $g(b^i, b^j)$.

$$g(b^i, b^j) \stackrel{(4.14)}{=} g(g^{ik}b_k, g^{j\ell}b_\ell) = g^{ik}g^{j\ell}g(b_k, b_\ell)$$
$$= g^{ik}g^{j\ell}g_{k\ell} \stackrel{(4.5)}{=} g^{ik}g^{j\ell}g_{\ell k} \stackrel{(4.13)}{=} g^{ik}\delta^j_k = g^{ij},$$

where we used in the second equality the bilinearity of g. Thus, generalizing Eq. (4.1), we have

$$\boxed{g^{ij} = g(b^i, b^j)}. \tag{4.16}$$

EXERCISE 4.24. In the setting of Exercise 4.21, verify Eq. (4.15) in the particular cases of $\mathcal{E}$ and $\mathcal{E}^g$ and of $\widetilde{\mathcal{B}}$ and $\widetilde{\mathcal{B}}^g$, that is verify that

(1) $\left(e^1\ e^2\ e^3\right) = \left(e_1\ e_2\ e_3\right) G^{-1}_{\mathcal{E}}$, and
(2) $\left(\tilde{b}^1\ \tilde{b}^2\ \tilde{b}^3\right) = \left(\tilde{b}_1\ \tilde{b}_2\ \tilde{b}_3\right) G^{-1}_{\widetilde{\mathcal{B}}}$.

Recall that, in Eq. (4.15), because of the way this equation was obtained, G is the matrix of g with respect to the basis $\mathcal{B}$.

Given that we just proved that reciprocal bases are unique, we can talk about *the* reciprocal basis (of a fixed vector space V associated to a basis and an inner product).

---

[3] Note that the following, like previously remarked, is a purely symbolic expression that has the only advantage of encoding the n expressions in Eq. (4.14) for $i = 1, \ldots, n$.

4.2 Reciprocal Basis

CLAIM 4.25 The reciprocal basis is **contravariant**.

Proof Let $\mathcal{B}$ and $\widetilde{\mathcal{B}}$ be two bases of V and $L := L_{\mathcal{B}\widetilde{\mathcal{B}}}$ be the corresponding matrix of the change of basis, with $\Lambda = L^{-1}$. Recall that this means that

$$\tilde{b}_i = L_i^j b_j \, .$$

We have to check that, if $\mathcal{B}^g = \{b^1, \ldots, b^n\}$ is a reciprocal basis for $\mathcal{B}$, then the basis $\{\tilde{b}^1, \ldots, \tilde{b}^n\}$ defined by

$$\tilde{b}^i = \Lambda_k^i b^k \tag{4.17}$$

is a reciprocal basis for $\widetilde{\mathcal{B}}$. Then the assertion will be proven, since $\{\tilde{b}^1, \ldots, \tilde{b}^n\}$ is contravariant by construction.

To check that $\{\tilde{b}^1, \ldots, \tilde{b}^n\}$ is the reciprocal basis for $\widetilde{\mathcal{B}}$, we need to check that, with the choice of $\tilde{b}^i$ as in Eq. (4.17), the property of the reciprocal basis given in Eq. (4.9) is verified, namely that

$$g(\tilde{b}^i, \tilde{b}_j) = \delta_j^i \, .$$

Indeed, we have

$$g(\tilde{b}^i, \tilde{b}_j) \stackrel{(4.17)}{=} g(\Lambda_k^i b^k, L_j^\ell b_\ell) = \Lambda_k^i L_j^\ell g(b^k, b_\ell) \stackrel{(4.9)}{=} \Lambda_k^i L_j^\ell \delta_\ell^k = \Lambda_k^i L_j^k = \delta_j^i \, ,$$

where the second equality comes from the bilinearity of g, and the last equality from the fact that $\Lambda = L^{-1}$. □

Suppose now that V is a vector space with a basis $\mathcal{B}$ and that $\mathcal{B}^g$ is the reciprocal basis of V with respect to $\mathcal{B}$ and to a fixed inner product $g : V \times V \to \mathbb{R}$. Then there are two ways of writing a vector $v \in V$, namely

$$v = \underbrace{v^i b_i}_{\text{with respect to } \mathcal{B}} = \underbrace{v_j b^j}_{\text{with respect to } \mathcal{B}^g} \, .$$

Recall that the (ordinary) coordinates of v with respect to $\mathcal{B}$ are *contravariant* (see Example 1.2).

CLAIM 4.26 Vector coordinates with respect to the reciprocal basis are **covariant**.

Proof This will follow from the fact that the reciprocal basis is contravariant and the idea of the proof is the same as in Claim 4.25.

Namely, let $\mathcal{B}, \widetilde{\mathcal{B}}$ be two bases of V, $L := L_{\mathcal{B}\widetilde{\mathcal{B}}}$ the matrix of the change of basis and $\Lambda = L^{-1}$. Let $\mathcal{B}^g$ and $\widetilde{\mathcal{B}}^g$ be the corresponding reciprocal bases and $v = v_j b^j$ a vector with respect to $\mathcal{B}^g$.

It is enough to check that the numbers

$$\tilde{v}_i := L_i^j v_j$$

are the coordinates of v with respect to $\tilde{\mathcal{B}}^g$, because in fact these coordinates are covariant by definition. But in fact, using this and Eq. (4.17), we obtain

$$\tilde{v}_i \tilde{b}^i = (L_i^j v_j)(\Lambda_k^i b^k) = \underbrace{L_i^j \Lambda_k^i}_{\delta_k^j} v_j b^k = v_j b^j = v$$

□

DEFINITION 4.27. The coordinates v_i of a vector $v \in V$ with respect to the reciprocal basis $\mathcal{B}^g$ are called the **covariant coordinates** of v.

Because of this covariant character, we represent the vector assembling the covariant coordinates as a *row* vector:

$$[v]_{\mathcal{B}^g} = \begin{pmatrix} v_1 & \ldots & v_n \end{pmatrix}.$$

4.2.3 Change of Basis from a Basis $\mathcal{B}$ to Its Reciprocal Basis $\mathcal{B}^g$

We want to look now at the direct relationship between the covariant and the contravariant coordinates of a vector v. From the alternative expressions

$$\underbrace{v^i b_i}_{\text{with respect to } \mathcal{B}} = v = \underbrace{v_j b^j}_{\text{with respect to } \mathcal{B}^g},$$

we obtain

$$(v^i g_{ij}) b^j = v^i (g_{ij} b^j) = v^i b_i = v = v_j b^j = (v_j g^{ji}) b_i$$

and hence, by comparing the coefficients of b^j,

$$\boxed{v_j = v^i g_{ij}} \quad \text{or} \quad \boxed{{}^t[v]_{\mathcal{B}^g} = G[v]_{\mathcal{B}}}, \tag{4.18}$$

Likewise, from

$$v^i b_i = v = v_j b^j = v_j (g^{ji} b_i) = (v_j g^{ji}) b_i ,$$

4.2 Reciprocal Basis

Table 4.1 Covariance and contravariance of vector coordinates

	$\mathcal{B} = \{b_1, \ldots, b_n\}$ basis		$\mathcal{B}^g = \{b^1, \ldots, b^n\}$ reciprocal basis
These bases are related by		$g(b^i, b_j) = \delta^i_j$	
A vector v has now two sets of coordinates	$v = v^i b_i$		$v = v_i b^i$
	contravariant coordinates		covariant coordinates
The matrices of g are	$g_{ij} = g(b_i, b_j)$		$g^{ij} = g(b^i, b^j)$
These matrices are inverse of each other, that is,		$g^{ik} g_{kj} = \delta^i_j$	
The basis and the reciprocal basis satisfy	$b_i = g_{ij} b^j$		$b^i = g^{ij} b_j$
Covariant and contravariant coordinates are related by	$v^i = g^{ij} v_j$		$v_i = g_{ij} v^j$

follows that (Table 4.1)

$$v^i = v_j g^{ji} \quad \text{or} \quad [v]_\mathcal{B} = G^{-1\,\mathrm{t}}[v]_{\mathcal{B}^g}. \tag{4.19}$$

Example 4.28 Let $\mathcal{E} = \{e_1, e_2, e_3\}$ be the standard basis of $\mathbb{R}^3$ and let

$$G = \begin{bmatrix} 1 & -1 & 0 \\ -1 & 2 & -1 \\ 0 & -1 & 2 \end{bmatrix}$$

be the matrix of g with respect to $\mathcal{E}$. In Eq. (4.10) of Example 4.20, we saw that

$$\mathcal{E}^g = \left\{ b^1 = \begin{bmatrix} 3 \\ 2 \\ 1 \end{bmatrix}, b^2 = \begin{bmatrix} 2 \\ 2 \\ 1 \end{bmatrix}, b^3 = \begin{bmatrix} 1 \\ 1 \\ 1 \end{bmatrix} \right\}$$

is the reciprocal basis of $\mathcal{E}$. We find the covariant coordinates of $v = \begin{bmatrix} 4 \\ 5 \\ 6 \end{bmatrix}$ with respect to $\mathcal{E}^g$ using Eq. (4.18), namely

$$^\mathrm{t}[v]_{\mathcal{E}^g} = G[v]_\mathcal{E} = \begin{bmatrix} 1 & -1 & 0 \\ -1 & 2 & -1 \\ 0 & -1 & 2 \end{bmatrix} \begin{pmatrix} 4 \\ 5 \\ 6 \end{pmatrix} = \begin{pmatrix} -1 \\ 0 \\ 7 \end{pmatrix}.$$

The following computation double checks this result:

$$v_i b^i = (-1)\begin{bmatrix} 3 \\ 2 \\ 1 \end{bmatrix} + 0 \begin{bmatrix} 2 \\ 2 \\ 1 \end{bmatrix} + 7 \begin{bmatrix} 1 \\ 1 \\ 1 \end{bmatrix} = \begin{bmatrix} 4 \\ 5 \\ 6 \end{bmatrix} = v.$$

□

Example 4.29 Let $V := P_{\leq 1}(\mathbb{R})$ be the vector space of polynomials of degree ≤ 1 (that is, "linear" polynomials of the form $a + bx$). Let $g : V \times V \to \mathbb{R}$ be defined by

$$g(p, q) := \int_0^1 p(x)q(x)dx,$$

and let $\mathcal{B} := \{1, x\}$ be a basis of V. Determine:

(1) the matrix G;
(2) the matrix G^{-1};
(3) the reciprocal basis $\mathcal{B}^g$;
(4) the contravariant coordinates of $p(x) = 6x$ (that is the coordinates of $p(x)$ with respect to $\mathcal{B}$);
(5) the covariant coordinates of $p(x) = 6x$ (that is the coordinates of $p(x)$ with respect to $\mathcal{B}^g$).

(1) The matrix G has entries $g_{ij} = g(b_i, b_i)$, that is

$$g_{11} = g(b_1, b_1) = \int_0^1 (b_1)^2 dx = \int_0^1 dx = 1$$

$$g_{12} = g(b_1, b_2) = \int_0^1 b_1 b_2 dx = \int_0^1 x = \frac{1}{2}$$

$$g_{21} = g(b_2, b_1) = \int_0^1 b_2 b_1 dx = \frac{1}{2}$$

$$g_{22} = g(b_2, b_2) = \int_0^1 (b_2)^2 dx = \int_0^1 x^2 dx = \frac{1}{3},$$

so that

$$G = \begin{pmatrix} 1 & \frac{1}{2} \\ \frac{1}{2} & \frac{1}{3} \end{pmatrix}.$$

(2) Since $\det G = 1 \cdot \frac{1}{3} - \frac{1}{2} \cdot \frac{1}{2} = \frac{1}{12}$, then by using the formula for the inverse given in Eq. (2.7), we get

$$G^{-1} = \begin{pmatrix} 4 & -6 \\ -6 & 12 \end{pmatrix}.$$

(3) Using Eq. (4.15), we obtain that

$$\begin{pmatrix} b^1 & b^2 \end{pmatrix} = \begin{pmatrix} 1 & x \end{pmatrix} G^{-1} = \begin{pmatrix} 1 & x \end{pmatrix} \begin{pmatrix} 4 & -6 \\ -6 & 12 \end{pmatrix} = \begin{pmatrix} 4 - 6x & -6 + 12x \end{pmatrix},$$

so that $\mathcal{B}^g = \{4 - 6x, -6 + 12x\}$.

(4) $p(x) = 6x = 0 \cdot 1 + 6 \cdot x$, so that $p(x)$ has contravariant coordinates $[p(x)]_\mathcal{B} = \begin{pmatrix} 0 \\ 6 \end{pmatrix}$.

(5) From Eq. (4.18) it follows that if $v = p(x)$, then

$$\begin{pmatrix} v_1 \\ v_2 \end{pmatrix} = G \begin{pmatrix} v^1 \\ v^2 \end{pmatrix} = \begin{pmatrix} 1 & \frac{1}{2} \\ \frac{1}{2} & \frac{1}{3} \end{pmatrix} \begin{pmatrix} 0 \\ 6 \end{pmatrix} = \begin{pmatrix} 3 \\ 2 \end{pmatrix}.$$

We can check this result:

$$v_1 b^1 + v_2 b^2 = 3 \cdot (4 - 6x) + 2 \cdot (-6 + 12x) = 6x.$$

□

4.2.4 Isomorphisms Between a Vector Space and Its Dual

We saw already in Proposition 3.10, that if V is a vector space and V^* is its dual, then $\dim V = \dim V^*$. In particular, this means that V and V^* can be identified, once we choose a basis $\mathcal{B}$ of V and a basis $\mathcal{B}^*$ of V^*. In fact, the basis $\mathcal{B}^*$ of V^* is given once we choose the basis $\mathcal{B}$ of V, as the dual basis of V^* with respect to $\mathcal{B}$. Then there is the following correspondence:

$$v \in V \longleftrightarrow \alpha \in V^*,$$

exactly when v and α have the same coordinates, respectively with respect to $\mathcal{B}$ and $\mathcal{B}^*$. However, this correspondence depends on the choice of the basis $\mathcal{B}$ and hence *not canonical*.

The above correspondence is an *isomorphism*. Recall that an **isomorphism** between vectors spaces V and W is an invertible linear transformation, $T : V \to W$. This is often used as *identification* of *equivalence* between vector spaces, denoted $\cong$.

When V is endowed with an inner product, then there is a **canonical identification** of V with V^* that is, an identification that does not depend on the basis $\mathcal{B}$ of V. In fact, let $g : V \times V \to \mathbb{R}$ be an inner product and let $v \in V$. Then

$$g(v, \cdot) : V \longrightarrow \mathbb{R}$$
$$w \longmapsto g(v, w)$$

is a linear form and hence we have the following *canonical* identification given by the metric

$$\begin{aligned} V &\longleftrightarrow V^* \\ v &\longleftrightarrow v^* := g(v, \cdot). \end{aligned} \qquad (4.20)$$

Note that the isomorphism sends the zero vector to the linear form identically equal to zero, since $g(v, \cdot) \equiv 0$ if and only if $v = 0$ by positive definiteness of g.

So far, we have two bases of the vector space V, namely the basis $\mathcal{B}$ and the reciprocal basis $\mathcal{B}^g$ and we have also the dual basis of the dual vector space V^*. It turns out that, under the isomorphism (4.20), the reciprocal basis of V and the dual basis of V^* correspond to each other. This follows from the fact that, under the isomorphism (4.20), an element of the reciprocal basis b^i corresponds to the linear form $g(b^i, \cdot)$

$$b^i \longleftrightarrow g(b^i, \cdot)$$

and the linear form $g(b^i, \cdot) : V \to \mathbb{R}$ has the property that

$$g(b^i, b_j) = \delta^i_j.$$

We conclude that

$$g(b^i, \cdot) \equiv \beta^i.$$

We have thus shown that:

▶ Under the canonical identification between V and V^* the reciprocal basis of V corresponds to the dual basis of V^*.

4.2.5 Geometric Interpretation

Let $g : V \times V \to \mathbb{R}$ be an inner product and $\mathcal{B} = \{b_1, \ldots, b_n\}$ a basis of V with reciprocal basis $\mathcal{B}^g$. Let $v = v_i b^i \in V$ be a vector written in terms of its covariant coordinates (that is, the coordinates with respect to the reciprocal basis). Then

$$g(v, b_k) = g(v_i b^i, b_k) = v_i \underbrace{g(b^i, b_k)}_{\delta^i_k} = v_k,$$

so that the formula in (4.8) becomes

$$\text{proj}_{b_k} v = \frac{v_k}{g(b_k, b_k)} b_k.$$

If we assume that the elements of the basis $\mathcal{B} = \{b_1, \ldots, b_n\}$ are unit vectors, then this further simplifies to give

$$\text{proj}_{b_k} v = v_k b_k. \tag{4.21}$$

This equation shows the following:

Fact 4.30 The covariant coordinates of v give the orthogonal projection of v onto $b_1, \ldots, b_n$.

Likewise, the following holds basically by definition:

Fact 4.31 The contravariant coordinates of v give the "parallel" projection of v onto $b_1, \ldots, b_n$ (Fig. 4.3).

4.3 Relevance of Covariance and Contravariance

Why do we need or care for covariant and contravariant components?

4.3.1 Physical Relevance

Consider the following physical problem: Calculate the work performed by a force F on a particle to move the particle by a small displacement dx, in the Euclidean plane. The work performed should be *independent* of the choice of the coordinate system (i.e. choice of basis) used, that is, *invariant*. For the work to remain independent of choice of basis we will see that, if the components of the displacement change contravariantly, then the components of the force should change covariantly.

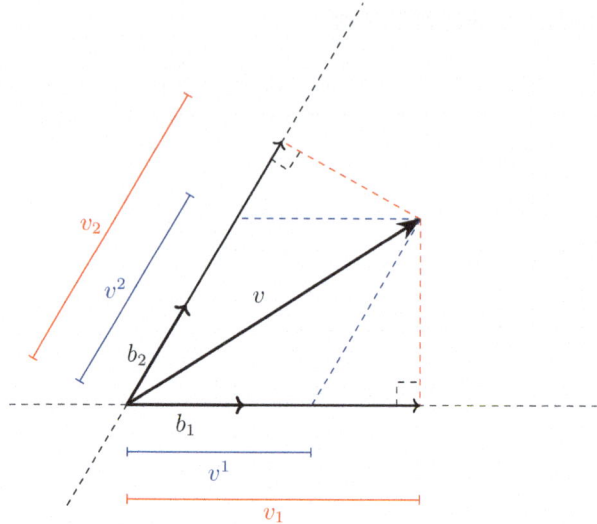

Fig. 4.3 Coordinates v^1, v^2 and covariant coordinates v_1, v_2 of vector v relative to basis $\mathcal{B} = \{b_1, b_2\}$

To see this let $\mathcal{B} = \{b_1, b_2\}$ be a basis of the Euclidean plane. Suppose the force $F = (F_1, F_2)$ is exerted on a particle that moves with a displacement $dx = (dx^1, dx^2)$. Then the work done is given by

$$dW = F_1 dx^1 + F_2 dx^2.$$

Suppose we are given another coordinate system $\widetilde{\mathcal{B}} := \{\tilde{b}_1, \tilde{b}_2\}$ and let $F = (\tilde{F}_1, \tilde{F}_2)$ and $dx = (d\tilde{x}^1, d\tilde{x}^2)$. Then

$$dW = \tilde{F}_1 d\tilde{x}^1 + \tilde{F}_2 d\tilde{x}^2.$$

Now assume that the coordinates of dx change contravariantly;

$$d\tilde{x}^i = \Lambda^i_j dx^j,$$

or, equivalently,

$$dx^i = L^i_j d\tilde{x}^j,$$

4.3 Relevance of Covariance and Contravariance

where $\Lambda = L^{-1}$ and $L = (L^i_j)$ is the change of basis matrix from $\mathcal{B}$ to $\tilde{\mathcal{B}}$.

$$\begin{aligned} dW &= F_1 dx^1 + F_2 dx^2 \\ &= F_1(L^1_1 d\tilde{x}^1 + L^1_2 d\tilde{x}^2) + F_2(L^2_1 d\tilde{x}^1 + L^2_2 d\tilde{x}^2) \\ &= (F_1 L^1_1 + F_2 L^2_1) d\tilde{x}^1 + (F_1 L^1_2 + F_2 L^2_2) d\tilde{x}^2 \end{aligned}$$

Since the work performed is independent of basis chosen, we also have

$$dW = \tilde{F}_1 d\tilde{x}^1 + \tilde{F}_2 d\tilde{x}^2.$$

This gives that

$$\tilde{F}_1 = F_1 L^1_1 + F_2 L^2_1 \quad \text{and} \quad \tilde{F}_2 = F_1 L^1_2 + F_2 L^2_2.$$

Hence the coordinates of F transform covariantly; $\tilde{F}_i = L^j_i F_j$. Using matrices this can be written as $(\tilde{F}_1, \tilde{F}_2) = (F_1, F_2)L$.

4.3.2 Starting Point

We start with the physical premise that physical scalar entities such as work (cf. Sect. 4.3.1) should be independent of choices of bases. Moreover, we declare displacement to be a vector, that is, a contravariant 1-tensor. Then it follows that position, velocity and acceleration are also contravariant 1-tensors; see the following exercise.

EXERCISE 4.32. Let T be a time-dependent vector. Verify that its time derivative, $\frac{dT}{dt}$ is also a vector. A similar result holds for tensors of other types.

The fact that, in a conservative force field, the force is the negative gradient of the potential energy function and the following exercise corroborate the covariance of force.

EXERCISE 4.33. Verify that the gradient of a real function on a vector space V is a covariant 1-tensor.

Therefore, in this text we set that:

- position or displacement, hence velocity and acceleration, are inherently contravariant objects;
- a gradient of a real function, hence a force, is inherently a covariant object.

On the other hand, Newton's second law states that *force is the time rate of change of momentum*:

$$F = \frac{dp}{dt}.$$

The *momentum p* is defined in terms of the Lagrangian function L governing the mechanics. The components of p in a certain given frame are the partial derivatives of L with respect to the components of the velocity. The velocity being a vector (hence contravariant), the momentum becomes a covector (hence covariant), fitting with the covariance of the force.

However, the momentum is often taken to be mass times velocity v. That turns out to be an important special case that occurs for constant mass and for a certain choice of the Lagrangian function L, namely $L = T - V$, where $T := \frac{m|v|^2}{2}$ is the kinetic energy, V is the potential energy and does not depend on velocity. Notice that, in this case, there is an inner product involved, in particular to measure the square length of v, and we work with orthonormal bases, hence the distinction between covariance and contravariance vanishes, as explained in Sect. 4.3.3.

The postulate that *the laws of Physics are the same in all inertial frames of reference* translates into the balance of equations regarding covariance and contravariance. This corresponds to a practical strategy for verifying the type of a tensor as illustrated in Sect. 4.3.1 and in the next exercise.

EXERCISE 4.34.

(1) Let T_{ij}, $i, j = 1, \ldots, n$, be the components of a $(0, 2)$-tensor T and u^i, $i = 1, \ldots, n$, the components of a $(1, 0)$-tensors u with respect to a basis $\mathcal{B}$ of an n-dimensional vector space. Show that then the numbers

$$v_j := T_{ij} u^i, \ j = 1, \ldots, n,$$

are the components of a $(0, 1)$-tensor v w.r.t. the basis $\mathcal{B}$.

(2) Now assume that, for each basis $\mathcal{B}$ of an n-dimensional vector space, we have numbers T_{ij}, $i, j = 1, \ldots, n$. We would like to find out, whether these numbers are the components of a $(0, 2)$-tensor T. Suppose we know, that, for each choice of a $(1, 0)$-tensor u with components u^i with respect to $\mathcal{B}$, the numbers

$$v_j := T_{ij} u^i, \ j = 1, \ldots, n,$$

are the components of a $(0, 1)$-tensor v. Show that then the T_{ij} are in fact the components of a $(0, 2)$-tensor.

Hint: It boils down to checking whether these numbers satisfy the relation $\tilde{T}_{ij} = L_i^{\ell_1} L_j^{\ell_2} T_{\ell_1 \ell_2}$ with respect to a change of basis L from $\mathcal{B}$ to $\tilde{\mathcal{B}}$.

4.3 Relevance of Covariance and Contravariance

In Physics, there is some leeway regarding the covariant or contravariant character when either some inner product is used (as addressed in Sect. 4.3.3), or the definition of the object is adjusted to the circumstances and changes character in this way (this could be the case of the different stress tensors in classical physics, continuum mechanics, electromagnetism and relativity). The above list may thus be seen as the convention of this text, justified by the considerations in Sect. 4.3.1.

4.3.3 Distinction Vanishes when Restricting to Orthonormal Bases

In the presence of an inner product $g : V \times V \to \mathbb{R}$ *and* if we restrict ourselves to orthonormal bases, the distinction between covariance and contravariance vanishes! In fact, it all becomes a matter of transposition: writing vectors as columns or as rows.

Why is that?

First, as we saw, the reciprocal of an orthonormal basis is equal to itself, so covariant and contravariant coordinates are equal in this special case.

Moreover, when we change from one orthonormal basis to another by a change of basis matrix L, the inverse change is given simply by the transpose of L. Here is a proof.

Let $\mathcal{B} = \{b_1, \ldots, b_n\}$ and $\widetilde{\mathcal{B}} = \{\tilde{b}_1, \ldots, \tilde{b}_n\}$ be two *orthonormal* bases of V and let $L := L_{\mathcal{B}\widetilde{\mathcal{B}}}$ be the corresponding matrix of the change of basis. This means, as always, that

$$\tilde{b}_i = L_i^j b_j \, .$$

Since $\mathcal{B}$ and $\widetilde{\mathcal{B}}$ are orthonormal, we have $g(b_i, b_j) = \delta_{ij} = g(\tilde{b}_i, \tilde{b}_j)$. Therefore, we have

$$\delta_{ij} = g(\tilde{b}_i, \tilde{b}_j) = g(L_i^k b_k, L_j^\ell b_\ell) = L_i^k L_j^\ell g(b_k, b_\ell) = L_i^k L_j^\ell \delta_{k\ell} = L_i^k L_j^k \, ,$$

showing that $L^{\mathsf{t}} L = I$, that is, $L^{-1} = {}^{\mathsf{t}} L$. Such a matrix L is called an **orthogonal matrix**.

We conclude that, in this special case, we see no distinction between covariant and contravariant behaviour, it is simply a matter of transposition.

Tensors 5

In this chapter, we address general tensors as a unifying tool for representing multilinear quantities across science. Moreover, we define tensor product in general and discuss some algebraic features of this product.

5.1 Towards General Tensors

Let V be a vector space. Up to now, we saw several objects related to V, which we called *tensors*. We summarize them in Table 5.1. From these, we infer the definition of a tensor of type $(0, q)$ for all $q \in \mathbb{N}$, but we cannot say the same for a tensor of type $(p, 0)$ for all $p \in \mathbb{N}$, even less of general type (p, q). The next discussion will lead us to tensors of type $(p, 0)$, and in the meantime we will discuss an important isomorphism.

5.1.1 Canonical Isomorphism Between V and $(V^*)^*$

We saw in Sect. 4.2.4 that any vector space is isomorphic to its dual, though in general the isomorphism is *not* canonical, that is, it depends on the choice of a basis. We also saw that, if there is an inner product on V, then there is a canonical isomorphism. The point of this section is to show that, even without an inner product, there is always a *canonical* isomorphism between V and its **bidual** $(V^*)^*$, that is the dual of its dual.

To see this, let us observe first of all that

$$\dim V = \dim(V^*)^*. \tag{5.1}$$

Table 5.1 Covariance and contravariance of aforementioned tensors

Tensor	Components	Behavior under a change of basis	Type
Vector in V	v^i	Contravariant tensor	$(1,0)$
Linear form $V \to \mathbb{R}$	α_j	Covariant tensor	$(0,1)$
Linear transformation $V \to V$	A^i_j	Tensor of mixed character: contravariant and covariant	$(1,1)$
Bilinear form $V \times V \to \mathbb{R}$	B_{ij}	Covariant 2-tensor	$(0,2)$
k-linear form $V \times \cdots \times V \to \mathbb{R}$	$F_{i_1 i_2 \ldots i_k}$	Covariant k-tensor	$(0,k)$

In fact, for any vector space W, we saw in Proposition 3.10 that $\dim W = \dim W^*$. If we apply this equality both to $W = V$ and to $W = V^*$, we obtain

$$\dim V = \dim V^* \quad \text{and} \quad \dim V^* = \dim(V^*)^*,$$

from which Eq. (5.1) follows immediately. We deduce (following Sect. 4.2.4) that V and $(V^*)^*$ are isomorphic, and we only have to see that there is a *canonical* isomorphism.

To this end, observe that a vector $v \in V$ gives rise to a linear form on V^* defined by

$$\psi_v : V^* \longrightarrow \mathbb{R}$$
$$\alpha \longmapsto \alpha(v).$$

Then we can define a linear map as follows:

$$\Psi : V \longrightarrow (V^*)^*$$
$$v \longmapsto \psi_v. \qquad (5.2)$$

Since, for any linear map $T : V \to W$ between vector spaces, we have the dimension formula (known as the *Rank-Nullity Theorem in Linear Algebra*):

$$\dim V = \dim \operatorname{im}(T) + \dim \ker(T),$$

it will be enough to show that $\ker \Psi = \{0\}$. In fact, in this case, we have

$$\dim V = \dim \operatorname{im}(\Psi),$$

and, since $\dim V = \dim(V^*)^*$, we can conclude that the subspace $\operatorname{im}(\Psi)$ must equal the whole space $(V^*)^*$. Hence, Ψ is an isomorphism. Notice that we have *not* chosen a basis to define the isomorphism Ψ.

To see that $\ker \Psi = \{0\}$, observe that this kernel consists of all vectors $v \in V$ such that $\alpha(v) = 0$ for all $\alpha \in V^*$. We want to see that the only vector $v \in V$ for which this happens is the zero vector. In fact, if $v \in V$ is non-zero and $\mathcal{B} =$

5.1 Towards General Tensors

$\{b_1, \ldots, b_n\}$ is *any* basis of V, then we can write $v = v^i b_i$, where at least one coordinate, say v^j, is not zero. In that case, if $\mathcal{B}^* = \{\beta_1, \ldots, \beta_n\}$ is the dual basis, we have $\beta_j(v) = v^j \neq 0$, thus we have found an element in V^* not vanishing on this v. We record this fact as follows:

Fact 5.1 Let V be a vector space and V^* its dual. The dual $(V^*)^*$ of V^* is canonically isomorphic to V. The canonical isomorphism $\Psi : V \to (V^*)^*$ takes $v \in V$ to the linear form on the dual $\psi_v : V^* \to \mathbb{R}$, $\psi_v(\alpha) := \alpha(v)$.

5.1.2 (2, 0)-Tensors

Recall that the dual of a vector space V is the vector space

$$V^* := \{\text{linear forms } \alpha : V \to \mathbb{R}\} = \{(0, 1)\text{-tensors}\}.$$

Taking now the dual of the vector space V^*, we obtain

$$(V^*)^* := \{\text{linear forms } \psi : V^* \to \mathbb{R}\}.$$

Using the canonical isomorphism $\Psi : V \to (V^*)^*$ and the fact that coordinate vectors are contravariant, we conclude that

$$\{\text{linear forms } \psi : V^* \to \mathbb{R}\} = (V^*)^* \cong V = \{(1, 0)\text{-tensors}\}.$$

So, changing the vector space from V to its dual V^* seems to have had the effect of converting covariant tensors of type $(0, 1)$ into contravariant ones of type $(1, 0)$.

We are going to apply the above principle to convert covariant tensors of type $(0, 2)$ into contravariant ones of type $(2, 0)$. Recall that

$$\{(0, 2)\text{-tensors}\} = \{\text{bilinear maps } \varphi : V \times V \to \mathbb{R}\}$$

and consider now

$$\{\text{bilinear maps } \sigma : V^* \times V^* \to \mathbb{R}\}.$$

Anticipating the contravariant character of such bilinear maps (to be proven in Sect. 5.1.4), we advance the following definition:

DEFINITION 5.2. A **tensor of type** $(2, 0)$ is a bilinear form on V^*, that is, a bilinear function $\sigma : V^* \times V^* \to \mathbb{R}$.

Then we have

$$\{(2,0)\text{-tensors}\} = \{\text{bilinear maps } \sigma : V^* \times V^* \to \mathbb{R}\}$$

and we denote this set $\mathrm{Bil}(V^* \times V^*, \mathbb{R})$.

EXERCISE 5.3. Check that $\mathrm{Bil}(V^* \times V^*, \mathbb{R})$ is a vector space. Just like in the case of $\mathrm{Bil}(V \times V, \mathbb{R})$ (cf. Exercise 3.26), it is enough to show that the zero function is in $\mathrm{Bil}(V^* \times V^*, \mathbb{R})$ and that if $\sigma, \tau \in \mathrm{Bil}(V^* \times V^*, \mathbb{R})$ and $c, d \in \mathbb{R}$, then the linear combination $c\sigma + d\tau$ is also in $\mathrm{Bil}(V^* \times V^*, \mathbb{R})$.

5.1.3 Tensor Product of Two Linear Forms on V^*

If $v, w \in V$ are two vectors (i.e., are two $(1, 0)$-tensors), we define

$$\sigma_{v,w} : V^* \times V^* \to \mathbb{R}$$

by

$$\sigma_{v,w}(\alpha, \beta) := \alpha(v)\beta(w),$$

for any two linear forms $\alpha, \beta \in V^*$. Then $\sigma_{v,w}$ is indeed *bilinear*, i.e., linear in each variable, α and β, so is indeed a $(2, 0)$-**tensor**. We denote

$$\boxed{\sigma_{v,w} =: v \otimes w}$$

and call this the **tensor product** of v and w.

Note 5.4 In general, we have

$$\boxed{v \otimes w \neq w \otimes v},$$

as there can be linear forms α, β such that $\alpha(v)\beta(w) \neq \alpha(w)\beta(v)$.

Similar to what we saw in Sect. 3.2.3, we find a basis for the space of $(2, 0)$-tensors by considering the $(2, 0)$-tensors defined by $b_i \otimes b_j$, where $\mathcal{B} = \{b_1, \ldots, b_n\}$ is a basis of V.

Proposition 5.5 *The elements* $b_i \otimes b_j$, $i, j = 1, \ldots, n$ *form a basis of* $\mathrm{Bil}(V^* \times V^*, \mathbb{R})$. *Thus* $\dim \mathrm{Bil}(V^* \times V^*, \mathbb{R}) = n^2$.

5.1 Towards General Tensors

The proof of this proposition is analogous to the one of Proposition 3.27 and we will not write it here. However, we state the crucial remark for the proof, analogous to Remark 3.28.

Remark 5.6 As for linear forms and bilinear forms on V, in order to verify that two bilinear forms on V^* are the same, it is enough to verify that they are the same on every pair of elements of a basis of V^*. In fact, let σ, τ be two bilinear forms on V^*, let $\{\gamma^1, \ldots, \gamma^n\}$ be a basis of V^* and let us assume that

$$\sigma(\gamma^i, \gamma^j) = \tau(\gamma^i, \gamma^j)$$

for all $1 \leq i, j, \leq n$. Let $\alpha = \alpha_i \gamma^i$ and $\beta = \beta_j \gamma^j$ be arbitrary elements of V^*. We now verify that $\sigma(\alpha, \beta) = \tau(\alpha, \beta)$. Because of the linearity in each variable, we have

$$\sigma(\alpha, \beta) = \sigma(\alpha_i \gamma^i, \beta_j \gamma^j) = \alpha_i \beta_j \sigma(\gamma^i, \gamma^j) = \alpha_i \beta_j \tau(\gamma^i, \gamma^j)$$
$$= \tau(\alpha_i \gamma^i, \beta_j \gamma^j) = \tau(\alpha, \beta).$$

$\square$

5.1.4 Contravariance of (2, 0)-Tensors

Let $\sigma : V^* \times V^* \to \mathbb{R}$ be a bilinear form on V^*, that is, a $(2, 0)$-tensor. We want to verify that it behaves as we expect with respect to a change of basis. After choosing a basis $\mathcal{B} = \{b_1, \ldots, b_n\}$ of V, we have the dual basis $\mathcal{B}^* = \{\beta^1, \ldots, \beta^n\}$ of V^* and the basis $\{b_i \otimes b_j : i, j = 1, \ldots, n\}$ of the space of $(2, 0)$-tensors.

The $(2, 0)$-tensor σ is represented by its components

$$\boxed{S^{ij} = \sigma(\beta^i, \beta^j)},$$

in the sense that

$$\boxed{\sigma = S^{ij} b_i \otimes b_j},$$

and the components S^{ij} can be arranged into a matrix[1]

$$S = \begin{pmatrix} S^{11} & \ldots & S^{1n} \\ \vdots & \ddots & \vdots \\ S^{n1} & \ldots & S^{nn} \end{pmatrix}$$

called the **matrix of the (2, 0)-tensor** σ with respect to the chosen basis of V.

---

[1] Once again, contrary to the change of basis matrix L, here we have only upper indices. This reflects the contravariance of the underlying tensor, σ.

We look now at how the components of a (2, 0)-tensor change with a change of basis. Let $\mathcal{B} = \{b_1, \ldots, b_n\}$ and $\widetilde{\mathcal{B}} = \{\tilde{b}_1, \ldots, \tilde{b}_n\}$ be two basis of V and let $\mathcal{B}^* := \{\beta^1, \ldots, \beta^n\}$ and $\widetilde{\mathcal{B}}^* := \{\tilde{\beta}^1, \ldots, \tilde{\beta}^n\}$ be the corresponding dual bases of V^*. Let $\sigma : V^* \times V^* \to \mathbb{R}$ be a (2, 0)-tensor with components

$$S^{ij} = \sigma(\beta^i, \beta^j) \quad \text{and} \quad \widetilde{S}^{ij} = \sigma(\tilde{\beta}^i, \tilde{\beta}^j)$$

with respect to $\mathcal{B}^*$ and $\widetilde{\mathcal{B}}^*$, respectively. Let $L := L_{\mathcal{B}\widetilde{\mathcal{B}}}$ be the matrix of the change of basis from $\mathcal{B}$ to $\widetilde{\mathcal{B}}$, and let $\Lambda := L^{-1}$. Then, as seen in Eqs. (2.4) and (3.11), we have that

$$\tilde{b}_j = L^i_j b_i \quad \text{and} \quad \tilde{\beta}^i = \Lambda^i_j \beta^j .$$

It follows that

$$\widetilde{S}^{ij} = \sigma(\tilde{\beta}^i, \tilde{\beta}^j) = \sigma(\Lambda^i_k \beta^k, \Lambda^j_\ell \beta^\ell) = \Lambda^i_k \Lambda^j_\ell \sigma(\beta^k, \beta^\ell) = \Lambda^i_k \Lambda^j_\ell S^{k\ell},$$

where the first and the last equalities follow from the definition of $\widetilde{S}^{ij}$ and of $S^{k\ell}$, respectively, the second from the change of basis and the third from the bilinearity of σ. We conclude that

$$\boxed{\widetilde{S}^{ij} = \Lambda^i_k \Lambda^j_\ell S^{k\ell}}. \tag{5.3}$$

Hence, the bilinear form σ is a **contravariant** 2-tensor.

EXERCISE 5.7. Verify that, in terms of matrices Eq. (5.3) translates into

$$\boxed{\widetilde{S} = \Lambda S^{\mathrm{t}} \Lambda}.$$

Compare with Eq. (3.15).

5.2 Tensors of Type (p, q)

In general, a (p, q)-tensor (with $p, q = 0, 1, 2, \ldots$) is defined to be a real-valued function of p covectors and of q vectors, which is linear in each of its arguments:

> DEFINITION 5.8. A **tensor of type** (p, q) or (p, q)**-tensor** is a multilinear form (or $(p + q)$-linear function)
>
> $$T : \underbrace{V^* \times \ldots V^*}_{p} \times \underbrace{V \times \cdots \times V}_{q} \longrightarrow \mathbb{R} .$$
>
> By convention, a tensor of type (0, 0) is a real number, a.k.a. *scalar* (a constant function of no arguments).

5.2 Tensors of Type (p, q)

Table 5.2 Aforementioned tensors viewed within general definition

Earlier tensor	Viewed as multilinear function	Type
Vector $v \in V$	$V^* \longrightarrow \mathbb{R}$ $\beta \longmapsto \beta(v)$	$(1, 0)$
Linear form $\alpha \in V^*$	$V \longrightarrow \mathbb{R}$ $w \longmapsto \alpha(w)$	$(0, 1)$
Linear transformation $F: V \to V$	$V^* \times V \longrightarrow \mathbb{R}$ $(\beta, w) \longmapsto \beta(F(w))$	$(1, 1)$
Bilinear form $\varphi \in \mathrm{Bil}(V \times V, \mathbb{R})$	$V \times V \longrightarrow \mathbb{R}$ $(v, w) \longmapsto \varphi(v, w)$	$(0, 2)$
k-linear form φ on V	$V \times \cdots \times V \longrightarrow \mathbb{R}$ $(v_1, \ldots, v_k) \longmapsto \varphi(v_1, \ldots, v_k)$	$(0, k)$
Bilinear form on V^* $\sigma \in \mathrm{Bil}(V^* \times V^*, \mathbb{R})$	$V^* \times V^* \longrightarrow \mathbb{R}$ $(\alpha, \beta) \longmapsto \sigma(\alpha, \beta)$	$(2, 0)$

The **order** of a tensor is the number of arguments that it takes: a tensor of type (p, q) has, thus, order $p + q$.[2]

If all the arguments of a tensor are vectors, i.e. $p = 0$, the tensor is said to be (purely) **covariant**. If the arguments are all linear forms, i.e. $q = 0$, the tensor is said to be (purely) **contravariant**. Otherwise, a (p, q)-tensor is of mixed character, p being its order of contravariance and q its order of covariance. Purely covariant tensors are what we earlier called multilinear forms. Purely contravariant tensors are sometimes called *polyadics*.[3]

Table 5.2 gives an overview of how the earlier examples of tensors fit in the above general definition.

Let T be a (p, q)-tensor, $\mathcal{B} = \{b_1, \ldots, b_n\}$ a basis of V and $\mathcal{B}^* = \{\beta^1, \ldots, \beta^n\}$ the corresponding dual basis of V^*. The **components** of T with respect to these bases are

$$\boxed{T^{i_1, \ldots, i_p}_{j_1, \ldots, j_q} := T(\beta^{i_1}, \ldots, \beta^{i_p}, b_{j_1}, \ldots, b_{j_q})}.$$

If, moreover, $\widetilde{\mathcal{B}} = \{\tilde{b}_1, \ldots, \tilde{b}_n\}$ is another basis, $\widetilde{\mathcal{B}}^* = \{\tilde{\beta}^1, \ldots, \tilde{\beta}^n\}$ the corresponding dual basis of V^* and $L := L_{\mathcal{B}\widetilde{\mathcal{B}}}$ the matrix of the change of basis with inverse

---

[2] The *order* of a tensor is sometimes also called *rank*. However, *rank* of a tensor is often reserved for another notion closer to the notion of *rank* of a matrix and related to *decomposability* of tensors (see Sects. 5.4.1 and 5.4.2).

[3] Whereas Latin roots are used for covariant tensors, like in *bi*linear form, Greek roots are used for contravariant tensors, like in *dy*adic, as established by Gibbs in late nineteen century.

$\Lambda := L^{-1}$, then the components of T with respect to these new bases are

$$\widetilde{T}^{i_1,\ldots,i_p}_{j_1,\ldots,j_q} = \Lambda^{i_1}_{k_1} \ldots \Lambda^{i_p}_{k_p} L^{\ell_1}_{j_1} \ldots L^{\ell_q}_{j_q} T^{k_1,\ldots,k_p}_{\ell_1,\ldots,\ell_q}.$$

The above formula displays the p-fold contravariant character and the q-fold covariant character of T.

The set of all tensors of type (p, q) on a vector space V with the natural operations of addition and scalar multiplication on tensors is itself a vector space denoted by

$$\mathcal{T}^p_q(V) := \{\text{all } (p, q)\text{-tensors on } V\}.$$

The **zero tensor** is *the* tensor all of whose components vanish, and two tensors are equal exactly when all their components are equal (with respect to any given basis).

EXERCISE 5.9. Let $V := \mathbb{R}^2$. Let T be the tensor of type $(3, 0)$ given with respect to the standard basis $\{\varepsilon^1, \varepsilon^2\}$ of V^* by

$$T(\varepsilon^i, \varepsilon^j, \varepsilon^k) := \begin{cases} i + i, & \text{when } k = 1 \\ i - j, & \text{when } k = 2. \end{cases}$$

(1) Compute $T((-1\ 2), (3\ 2), (1\ 1))$.
(2) Determine the components of T with respect to the basis

$$\mathcal{B} := \left\{ \begin{bmatrix} 1 \\ -1 \end{bmatrix}, \begin{bmatrix} 1 \\ 0 \end{bmatrix} \right\}$$

(and the corresponding dual basis $\mathcal{B}^*$).

5.3 Symmetric and Antisymmetric Tensors

We consider now (purely) *covariant* q-tensors on the vector space V, i.e., real-valued functions of the form

$$T : \underbrace{V \times \cdots \times V}_{q} \longrightarrow \mathbb{R}$$

which are linear in each of its q arguments. Similar notions hold for (purely) *contravariant* tensors.

5.3 Symmetric and Antisymmetric Tensors

> **DEFINITION 5.10.** A $(0, q)$-tensor T is (totally) **symmetric**, if the result $T(v_1, \ldots, v_q)$ is independent of the order of the arguments, for all $v_1, \ldots, v_q \in V$.
>
> A $(0, q)$-tensor T is called (totally) **antisymmetric** or **skewsymmetric**, if the result $T(v_1, \ldots, v_q)$ changes sign every time we swap two of its arguments, for all $v_1, \ldots, v_q \in V$.

For the special case of a covariant 2-tensor, that is, of a bilinear form $T : V \times V \to \mathbb{R}$, we say that T is *symmetric* if

$$T(v_1, v_2) = T(v_2, v_1), \qquad \text{for all } v_1, v_2 \in V,$$

and that T is *antisymmetric* if

$$T(v_1, v_2) = -T(v_2, v_1), \qquad \text{for all } v_1, v_2 \in V.$$

Inner products are examples of symmetric $(0, 2)$-tensors.

> **EXERCISE 5.11.** Check that the set of all symmetric covariant k-tensors, denoted $S^k V^*$, is a real subspace of the space of all covariant k-tensors, $\mathcal{T}_k^0(V)$.
>
> Similarly, the set of all antisymmetric covariant k-tensors, denoted $\bigwedge^k V^*$, is a real subspace of $\mathcal{T}_k^0(V)$.

> **EXERCISE 5.12.** We consider the spaces of the previous exercise with $k = 2$. Let n be the dimension of the vector space V. Show that
>
> $$\dim S^2 V^* = \frac{n(n+1)}{2} \quad \text{and} \quad \dim \bigwedge^2 V^* = \frac{n(n-1)}{2}.$$
>
> *Hint:* Show that, if $\beta^1, \ldots, \beta^n$ form a basis of V^*, then
>
> $$\left\{ \tfrac{1}{2} \left(\beta^i \otimes \beta^j + \beta^j \otimes \beta^i \right) \mid i \geq j \right\}$$
>
> is a basis of $S^2 V^*$ and
>
> $$\left\{ \tfrac{1}{2} \left(\beta^i \otimes \beta^j - \beta^j \otimes \beta^i \right) \mid i > j \right\}$$
>
> is a basis of $\bigwedge^2 V^*$. Then count the number of elements in these bases.

5.4 Tensor Product

We saw already in Sects. 3.2.3 and 3.3.3 the tensor product of two multilinear forms. Since multilinear forms are covariant tensors, we said that this corresponds to the tensor product of two covariant tensors. We can now define the tensor product of two tensors in general. This will further lead us to the tensor product of vector spaces.

5.4.1 Tensor Product of Tensors

DEFINITION 5.13. Let

$$T: \underbrace{V^* \times \cdots \times V^*}_{p} \times \underbrace{V \times \cdots \times V}_{q} \longrightarrow \mathbb{R}$$

be a (p, q)-tensor and

$$U: \underbrace{V^* \times \cdots \times V^*}_{k} \times \underbrace{V \times \cdots \times V}_{\ell} \longrightarrow \mathbb{R}$$

a (k, ℓ) tensor. The **tensor product** $T \otimes U$ of T and U is the $(p+k, q+\ell)$-tensor

$$T \otimes U: \underbrace{V^* \times \cdots \times V^*}_{p+k} \times \underbrace{V \times \cdots \times V}_{q+\ell} \longrightarrow \mathbb{R}$$

defined by

$$(T \otimes U)(\alpha_1, \ldots, \alpha_{p+k}, v_1, \ldots, v_{q+\ell}) :=$$
$$T(\alpha_1, \ldots, \alpha_p, v_1, \ldots, v_q) U(\alpha_{p+1}, \ldots, \alpha_{p+k}, v_{q+1}, \ldots, v_{q+\ell}).$$

Although both $T \otimes U$ and $U \otimes T$ are tensors of the same type, in general we have

$$T \otimes U \neq U \otimes T.$$

So we say that the tensor product is *not commutative*. On the other hand, the tensor product is *associative*, since we always have

$$(S \otimes T) \otimes U = S \otimes (U \otimes T).$$

Analogously to how we proceeded in the case of $(0, 2)$-tensors, we compute the dimension of the vector space $\mathcal{T}_q^p(V)$. Let $\mathcal{B} = \{b_1, \ldots, b_n\}$ be a basis of V and $\mathcal{B}^* := \{\beta^1, \ldots, \beta^n\}$ the corresponding dual basis of V^*. Just like we saw in Proposition 5.5 in the case of $(0, 2)$-tensors, we form a basis of $\mathcal{T}_q^p(V)$ by collecting

5.4 Tensor Product

all elements of the form

$$b_{i_1} \otimes b_{i_2} \otimes \cdots \otimes b_{i_p} \otimes \beta^{j_1} \otimes \beta^{j_2} \otimes \cdots \otimes \beta^{j_q}$$

where the indices $i_1, \ldots, i_p$ and $j_1, \ldots, j_q$ take all values between 1 and n. Since there are $n^p n^q = n^{p+q}$ elements in the above basis (corresponding to all possible choices of b_{i_k} and β^{j_ℓ}), we deduce that

$$\boxed{\dim \mathcal{T}_q^p(V) = n^{p+q}}.$$

If T is a (p, q)-tensor with components

$$T^{i_1,\ldots,i_p}_{j_1,\ldots,j_q} := T(\beta^{i_1}, \ldots, \beta^{i_p}, b_{j_1}, \ldots, b_{j_q}),$$

then we have, with a $(p+q)$-fold application of the Einstein convention, that

$$T = T^{i_1,\ldots,i_p}_{j_1,\ldots,j_q} b_{i_1} \otimes b_{i_2} \otimes \cdots \otimes b_{i_p} \otimes \beta^{j_1} \otimes \beta^{j_2} \otimes \cdots \otimes \beta^{j_q}.$$

A **simple tensor** (also called a **tensor of rank 1** or **pure tensor** or **decomposable tensor** or **elementary tensor**) of type (p, q) is a tensor T that can be written as a tensor product of the form

$$T = \underbrace{a \otimes b \otimes \ldots}_{p} \otimes \underbrace{\alpha \otimes \beta \otimes \ldots}_{q}$$

where $a, b, \ldots \in V$ and $\alpha, \beta, \ldots \in V^*$. The **rank of a tensor** T is then the minimum number of simple tensors that sum to T.[4] By convention, the zero tensor has rank 0 and a non-zero tensor of order zero, i.e., a non-zero scalar, has rank 1. A non-zero tensor of order 1 always has rank 1. Already among tensors of order 2 (and when $\dim V \geq 2$) there are tensors of rank greater than 1. Example 5.15 provides such an instance.

5.4.2 Tensor Product for Vector Spaces

To complement the previous exposition and justify the notation $V^* \otimes V^*$ for the vector space of all bilinear forms on V (cf. Sect. 3.2.3), we aim in this section to give an idea of what the *tensor product* for finite-dimensional vector spaces should mean and of how the tensor product *for vector spaces* relates to the tensor product *for tensors*.

---

[4] This notion of *rank* of a tensor extends the notion of *rank* of a matrix, as can be seen by considering tensors of order two and their corresponding matrices of components.

Let V and W be two vector spaces with $\dim V = n$ and $\dim W = m$. Choose $\{b_1, \ldots, b_n\}$ a basis of V and $\{a_1, \ldots, a_m\}$ a basis of W.

DEFINITION 5.14. The **tensor product** of V and W is the $(n \cdot m)$-dimensional vector space $V \otimes W$ with basis

$$\{b_i \otimes a_j : 1 \leq i \leq n, \ 1 \leq j \leq m\}.$$

Elements of $V \otimes W$ are naturally referred to as *tensors*. They can be seen as bilinear maps $V^* \times W^* \to \mathbb{R}$, without depending on the choice of bases: By the above definition, tensors are linear combinations of the $b_i \otimes a_j$. By viewing $b_i \in V$ as a *linear map* (cf. Table 5.2)

$$b_i : V^* \longrightarrow \mathbb{R}, \qquad \beta \longmapsto \beta(b_i),$$

and similarly for $a_j \in W$ as

$$a_j : W^* \longrightarrow \mathbb{R}, \qquad \alpha \longmapsto \alpha(a_j),$$

we regard the symbol $b_i \otimes a_j$ as a bilinear map

$$V^* \times W^* \longrightarrow \mathbb{R}, \qquad (\beta, \alpha) \longmapsto \beta(b_i)\alpha(a_j).$$

The tensor product $V \otimes W$ is endowed with a bilinear map from the cartesian product $V \times W$

$$\Psi : V \times W \longrightarrow V \otimes W$$

defined as follows. If

$$v = v^i b_i \in V \qquad \text{and} \qquad w = w^j a_j \in W,$$

then $\Psi(v, w) =: v \otimes w$ is the element of $V \otimes W$ with coordinates $v^i w^j$ with respect to the basis $\{b_i \otimes a_j : 1 \leq i \leq n, \ 1 \leq j \leq m\}$, so that the following holds:

$$v \otimes w = \left(v^i b_i\right) \otimes \left(w^j a_j\right) = \left(v^i w^j\right) b_i \otimes a_j.$$

5.4 Tensor Product

Notice that the ranges for the indices are different: $1 \leq i \leq n$, $1 \leq j \leq m$. The numbers $v^i w^j$ may be viewed as obtained by the *outer product* of the coordinate vectors of v and w yielding an $n \times m$ matrix:

$$[v]_{\mathcal{B}}{}^t [w]_{\mathcal{A}} = \begin{pmatrix} v^1 \\ \vdots \\ v^n \end{pmatrix} \begin{pmatrix} w^1 & \ldots & w^m \end{pmatrix} = \begin{pmatrix} v^1 w^1 & \ldots & v^1 w^m \\ \vdots & & \vdots \\ v^n w^1 & \ldots & v^n w^m \end{pmatrix}.$$

An element of $V \otimes W$ that is in the image of the map Ψ, that is, an element of $V \otimes W$ that can be written as $v \otimes w$ for some $v \in V$ and $w \in W$ is called a **simple tensor** (or **tensor of rank 1** or **pure tensor** or **decomposable tensor** or **elementary tensor**). Yet keep in mind, that the map Ψ is usually by far *not* surjective. In particular, one can show that, if $v_1, v_2 \in V$ are linearly independent and $w_1, w_2 \in W$ are also linearly independent, then the sum $v_1 \otimes w_1 + v_2 \otimes w_2$ is not a pure tensor. Checking the first instance of this phenomenon is left as the next exercise. The proof in general goes along somewhat similar lines, but gets sophisticated.[5]

EXERCISE 5.15. Check that if both V and W are 2-dimensional with respective bases $\{b_1, b_2\}$ and $\{a_1, a_2\}$, then $b_1 \otimes a_1 + b_2 \otimes a_2$ is not a pure tensor.

The following proposition gives some useful identifications.

Proposition 5.16 *Let V and W be vector spaces with* $\dim V = n$ *and* $\dim W = m$ *and let*

$$\mathrm{Lin}(V, W^*) := \{\textit{linear maps } V \to W^*\}.$$

Then

$$\mathrm{Bil}(V \times W, \mathbb{R}) \cong \mathrm{Lin}(V, W^*)$$
$$\cong \mathrm{Lin}(W, V^*)$$
$$\cong V^* \otimes W^*$$
$$\cong (V \otimes W)^*$$
$$= \mathrm{Lin}(V \otimes W, \mathbb{R}).$$

Proof Here is the idea behind this chain of canonical isomorphisms. Let $f \in \mathrm{Bil}(V \times W, \mathbb{R})$, that is, $f : V \times W \to \mathbb{R}$ is a bilinear function, in particular it takes two vectors, $v \in V$ and $w \in W$, as input and gives a real number $f(v, w) \in \mathbb{R}$ as output. If, however, we only feed f one vector $v \in V$ as input, then there is a

---

[5] The *Segre embedding* from Algebraic Geometry provides the framework for understanding this properly.

remaining spot waiting for a vector $w \in W$ to produce a real number. Since f is linear in V and in W, the map $f(v, \cdot) : W \to \mathbb{R}$ is a linear form, so $f(v, \cdot) \in W^*$, hence f gives us an element in $\mathrm{Lin}(V, W^*)$. There is then a linear map

$$\mathrm{Bil}(V \times W, \mathbb{R}) \longrightarrow \mathrm{Lin}(V, W^*)$$
$$f \longmapsto T_f,$$

where

$$T_f(v)(w) := f(v, w).$$

Conversely, any $T \in \mathrm{Lin}(V, W^*)$ can be identified with a bilinear map $f_T \in \mathrm{Bil}(V \times W, \mathbb{R})$ defined by

$$f_T(v, w) := T(v)(w).$$

Since $f_{T_f} = f$ and $T_{f_T} = T$, we have proven the first isomorphism in the proposition.

Analogously, if the input is only a vector $w \in W$, then $f(\cdot, w) : V \to \mathbb{R}$ is a linear map and we now see that $f \in \mathrm{Bil}(V \times W, \mathbb{R})$ defines a linear map $U_f \in \mathrm{Lin}(W, V^*)$. The same reasoning as in the previous paragraph provides the canonical isomorphism $\mathrm{Bil}(V \times W, \mathbb{R}) \cong \mathrm{Lin}(W, V^*)$.

Observe now that, because of our definition of $V^* \otimes W^*$, we have

$$\mathrm{Bil}(V \times W, \mathbb{R}) \cong V^* \otimes W^*,$$

since these spaces both have basis

$$\{\beta^i \otimes \alpha^j : 1 \leq i \leq n,\ 1 \leq j \leq m\},$$

where $\{b_1, \ldots, b_n\}$ is a basis of V with corresponding dual basis $\{\beta^1, \ldots, \beta^n\}$ of V^*, and $\{a_1, \ldots, a_n\}$ is a basis of W with corresponding dual basis $\{\alpha^1, \ldots, \alpha^n\}$ of W^*.

Finally, an element $D_{ij}\beta^i \otimes \alpha^j \in V^* \otimes W^*$ may be viewed as a linear map $V \otimes W \to \mathbb{R}$, that is as an element of $(V \otimes W)^*$ by

$$V \otimes W \longrightarrow \mathbb{R}$$
$$C^{k\ell} b_k \otimes a_\ell \longmapsto D_{ij} C^{k\ell} \underbrace{\beta^i(b_k)}_{\delta^i_k} \underbrace{\alpha^j(a_\ell)}_{\delta^j_\ell} = D_{ij} C^{k\ell}.$$

$\square$

5.4 Tensor Product

Because of the identification $\text{Bil}(V \times W, \mathbb{R}) \cong \text{Lin}(V \otimes W, \mathbb{R})$, we can say that

> *the tensor product linearizes what was bilinear (or multilinear)*.

There is no reason to restrict oneself to the tensor product of only two factors. One can equally define the tensor product $V_1 \otimes \cdots \otimes V_k$, and obtain a vector space of dimension $\dim V_1 \times \cdots \times \dim V_k$. Note that we do not need to use brackets, since the tensor product is associative: $(V_1 \otimes V_2) \otimes V_3 = V_1 \otimes (V_2 \otimes V_3)$.

We have

$$\mathcal{T}^p_q(V) = \underbrace{V \otimes \cdots \otimes V}_{p} \otimes \underbrace{V^* \otimes \cdots \otimes V^*}_{q},$$

since both spaces have the same basis. An element T of $\mathcal{T}^p_q(V)$ was first regarded according to Definition 5.8 as a multilinear map

$$T : \underbrace{V^* \times \ldots V^*}_{p} \times \underbrace{V \times \cdots \times V}_{q} \longrightarrow \mathbb{R}.$$

Now, with respect to bases $\mathcal{B} = \{b_1, \ldots, b_n\}$ of V and $\mathcal{B}^* = \{\beta^1, \ldots, \beta^n\}$ of V^*, the components of T are

$$T^{i_1,\ldots,i_p}_{j_1,\ldots,j_q} := T(\beta^{i_1}, \ldots, \beta^{i_p}, b_{j_1}, \ldots, b_{j_q}),$$

hence we may view T as

$$T^{i_1,\ldots,i_p}_{j_1,\ldots,j_q} b_{i_1} \otimes \ldots \otimes b_{i_p} \otimes \beta^{j_1} \otimes \ldots \otimes \beta^{j_q} \in \underbrace{V \otimes \cdots \otimes V}_{p} \otimes \underbrace{V^* \otimes \cdots \otimes V^*}_{q}.$$

In particular, the pth **tensor power** of a vector space is

$$V^{\otimes p} := \underbrace{V \otimes \cdots \otimes V}_{p} = \mathcal{T}^p_0(V).$$

Therefore, we may write

$$\boxed{\text{Bil}(V^* \times V^*, \mathbb{R}) = V \otimes V = V^{\otimes 2}}, \quad \boxed{\text{Bil}(V \times V, \mathbb{R}) = V^* \otimes V^* = (V^*)^{\otimes 2}}$$

and

$$\boxed{\mathcal{T}^p_q(V) = V^{\otimes p} \otimes (V^*)^{\otimes q}}.$$

Some Physical Tensors 6

Mathematically-speaking, a tensor is a real-valued function of some number of vectors and some number of covectors (a.k.a. linear forms), which is linear in each of its arguments. On the other hand, tensors are most useful in connection with concrete physical applications.

In this chapter, we borrow notions and computations from Physics and Calculus to discuss important classical tensors. We assume familiarity with concepts from Physics, for instance at the level of [14] or [15] and from Calculus, for instance at the level of [6] or [12].

6.1 Inertia Tensor

6.1.1 Physical Preliminaries

We consider a rigid body M fixed at a point O and rotating about an axis through O with **angular velocity** ω. Denoting the time variable t and an angle variable θ around the axis of rotation, the angular velocity will be viewed as a vector[1] with magnitude

$$\|\omega\| = \left\| \frac{d\theta}{dt} \right\|,$$

---
[1] *Warning:* The angular velocity is actually only what physicists call a *pseudovector* because it does not follow the usual contravariance of a vector in case of orientation flip. Luckily, this issue does not affect the inertia tensor, since the sign flip cancels out thanks to squaring.

© The Author(s), under exclusive license to Springer Nature Switzerland AG 2025
A. Cannas da Silva et al., *Tensors for Scientists*, Compact Textbooks in Mathematics, https://doi.org/10.1007/978-3-031-94136-8_6

with direction given by the axis of rotation and with orientation given by the right-hand rule. The **position vector** of a point P in the body M relative to the origin O is

$$\mathbf{x} := \overrightarrow{OP}$$

while the **linear velocity** of that point P is

$$\mathbf{v} := \boldsymbol{\omega} \times \mathbf{x}.$$

The linear velocity $\mathbf{v}$ has, hence, magnitude

$$\|\mathbf{v}\| = \underbrace{\|\boldsymbol{\omega}\|}_{\left\|\frac{d\theta}{dt}\right\|} \underbrace{\|\mathbf{x}\| \sin \alpha}_{=:r},$$

where α is the angle between ω and $\mathbf{x}$, and has direction tangent at P to the circle of radius r perpendicular to the axis of rotation (Fig. 6.1).

The **kinetic energy** of an infinitesimal region dM of M around P is

$$dE = \frac{1}{2}\|\mathbf{v}\|^2 dm,$$

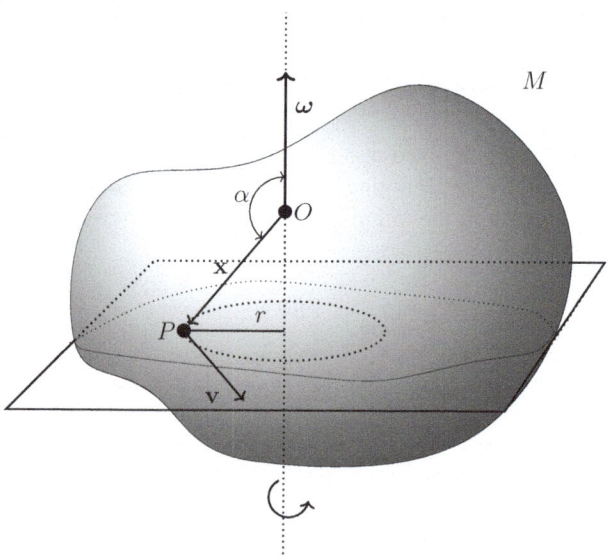

Fig. 6.1 Computing the kinetic energy of a rigid body rotating about an axis

6.1 Inertia Tensor

where $\|\mathbf{v}\|^2 = \mathbf{v} \cdot \mathbf{v}$ is the square of the norm of the linear velocity and dm is the mass of dM. The **total kinetic energy** of M is

$$E = \frac{1}{2} \int_M \|\mathbf{v}\|^2 dm = \frac{1}{2} \int_M \|\boldsymbol{\omega} \times \mathbf{x}\|^2 dm .$$

Actually, depending on the type of rigid body, we might take here a sum instead of integral, or some other type of integral (line integral, surface integral, etc) such as:

(1) If M is a solid in 3-dimensional space, then

$$E = \frac{1}{2} \iiint_M \|\boldsymbol{\omega} \times \mathbf{x}_P\|^2 \rho_P \, dx^1 dx^2 dx^3 ,$$

where the norm squared $\|\boldsymbol{\omega} \times \mathbf{x}_P\|^2$ and the density ρ_P are functions of the point P with coordinates (x^1, x^2, x^3).

(2) If M is a flat sheet, then

$$E = \frac{1}{2} \iint_M \|\boldsymbol{\omega} \times \mathbf{x}_P\|^2 \rho_P \, dx^1 dx^2 ,$$

where the integrand only depends on two cartesian coordinates.

(3) If M is a (curvy) surface in 3-dimensional space, then

$$E = \frac{1}{2} \iint_M \|\boldsymbol{\omega} \times \mathbf{x}_P\|^2 \rho_P \, d\sigma ,$$

where $d\sigma$ is the infinitesimal element of the surface for a surface integral.

(4) If M is a wire in 3-dimensional space, then

$$E = \frac{1}{2} \int_M \|\boldsymbol{\omega} \times \mathbf{x}_P\|^2 \rho_P \, ds ,$$

where ds is the infinitesimal element of length for a line integral.

(5) If M is a finite set of N point masses m_i with rigid relative positions, then

$$E = \frac{1}{2} \sum_{i=1}^{N} \|\boldsymbol{\omega} \times \mathbf{x_i}\|^2 m_i .$$

We will keep writing our formulas for the first case (with a volume integral); these should be adjusted for situations of the other types.

In any case, we need to work out the quantity

$$\|\boldsymbol{\omega} \times \mathbf{x}\|^2$$

for vectors $\boldsymbol{\omega}$ and $\mathbf{x}$ in 3-dimensional space.

To this purpose, we use the **Lagrange identity**,[2] according to which

$$(a \times b) \cdot (c \times d) = \det \begin{bmatrix} a \cdot c & a \cdot d \\ b \cdot c & b \cdot d \end{bmatrix}. \tag{6.1}$$

Applying the identity (6.1) with $a = c = \boldsymbol{\omega}$ and $b = d = \mathbf{x}$, we obtain

$$\|\boldsymbol{\omega} \times \mathbf{x}\|^2 = (\boldsymbol{\omega} \times \mathbf{x}) \cdot (\boldsymbol{\omega} \times \mathbf{x}) = \det \begin{bmatrix} \boldsymbol{\omega} \cdot \boldsymbol{\omega} & \boldsymbol{\omega} \cdot \mathbf{x} \\ \mathbf{x} \cdot \boldsymbol{\omega} & \mathbf{x} \cdot \mathbf{x} \end{bmatrix} = \|\boldsymbol{\omega}\|^2 \|\mathbf{x}\|^2 - \|\boldsymbol{\omega} \cdot \mathbf{x}\|^2.$$

Let now $\mathcal{B} = \{e_1, e_2, e_3\}$ be an orthonormal[3] basis of $\mathbb{R}^3$, so that

$$\boldsymbol{\omega} = \omega^i e_i \quad \text{and} \quad \mathbf{x} = x^i e_i.$$

Then

$$\|\boldsymbol{\omega}\|^2 = \boldsymbol{\omega} \cdot \boldsymbol{\omega} = \delta_{ij}\omega^i \omega^j = \omega^1 \omega^1 + \omega^2 \omega^2 + \omega^3 \omega^3$$

$$\|\mathbf{x}\|^2 = \mathbf{x} \cdot \mathbf{x} = \delta_{k\ell} x^k x^\ell = x^1 x^1 + x^2 x^2 + x^3 x^3$$

$$\boldsymbol{\omega} \cdot \mathbf{x} = \delta_{ik}\omega^i x^k = \omega^1 x^1 + \omega^2 x^2 + \omega^3 x^3$$

so that

$$\|\boldsymbol{\omega} \times \mathbf{x}\|^2 = \|\boldsymbol{\omega}\|^2 \|\mathbf{x}\|^2 - \|\boldsymbol{\omega} \cdot \mathbf{x}\|^2$$

$$= (\delta_{ij}\omega^i \omega^j)(\delta_{k\ell} x^k x^\ell) - (\delta_{ik}\omega^i x^k)(\delta_{j\ell}\omega^j x^\ell)$$

$$= (\delta_{ij}\delta_{k\ell} - \delta_{ik}\delta_{j\ell})\omega^i \omega^j x^k x^\ell.$$

Therefore, the total kinetic energy is

$$\boxed{E = \frac{1}{2}(\delta_{ij}\delta_{k\ell} - \delta_{ik}\delta_{j\ell})\omega^i \omega^j \iiint_M x^k x^\ell \, dm}$$

and it depends only on $\omega^1, \omega^2, \omega^3$ (since we have integrated over the x^1, x^2, x^3).

---

[2] The Lagrange identity can be patiently proven in coordinates.

[3] We could use any basis of $\mathbb{R}^3$. Then, instead of the δ_{ij}, the formulas would have involved the components of the metric tensor g_{ij}. However, computations with orthonormal bases are simpler; in particular, the inverse of an orthonormal basis change L is simply ${}^t L$. Moreover, the inertia tensor is symmetric, hence admits an orthonormal eigenbasis.

6.1.2 Moments of Inertia

> **DEFINITION 6.1.** The **inertia tensor** is the covariant 2-tensor whose components with respect to an orthonormal basis $\mathcal{B}$ are
>
> $$\boxed{I_{ij} = (\delta_{ij}\delta_{k\ell} - \delta_{ik}\delta_{j\ell}) \iiint_M x^k x^\ell \, dm}.$$

Then the kinetic energy of the rotating rigid body is

$$\boxed{E = \frac{1}{2} I_{ij} \omega^i \omega^j}$$

which in matrix notation amounts to

$$E = \frac{1}{2}\boldsymbol{\omega} \cdot I\boldsymbol{\omega} = \frac{1}{2}{}^{\mathrm{t}}\boldsymbol{\omega} I \boldsymbol{\omega}.$$

Remark 6.2 If, instead of an orthonormal basis, we had used any basis of $\mathbb{R}^3$, we would have gotten

$$I_{ij} = (g_{ij} g_{k\ell} - g_{ik} g_{j\ell}) \iiint_M x^k x^\ell \, dm.$$

where g_{ij} are the components of the metric tensor. This formula also makes apparent the covariance and the symmetry of I inherited from the metric: $I_{ij} = I_{ji}$ for all i and j.

We will see that the inertia tensor is a convenient way to encode all moments of inertia of an object in one quantity and we return now to the case of an orthonormal basis. The first component of the inertia tensor is

$$I_{11} = \underbrace{(\delta_{11}\delta_{k\ell}}_{\substack{=0 \\ \text{unless} \\ k=\ell}} - \underbrace{\delta_{1k}\delta_{1\ell})}_{\substack{=0 \\ \text{unless} \\ k=\ell=1}} \iiint_M x^k x^\ell \, dm.$$

If $k = \ell = 1$, then $\delta_{11}\delta_{11} - \delta_{11}\delta_{11} = 0$, so that the non-vanishing terms have $k = \ell \neq 1$. In this way, one can check that

$$I_{11} = \iiint_M (x^2 x^2 + x^3 x^3)\, dm$$

$$I_{22} = \iiint_M (x^1 x^1 + x^3 x^3)\, dm$$

$$I_{33} = \iiint_M (x^1 x^1 + x^2 x^2)\, dm$$

$$I_{23} = I_{32} = -\iiint_M x^2 x^3\, dm$$

$$I_{31} = I_{13} = -\iiint_M x^1 x^3\, dm$$

$$I_{12} = I_{21} = -\iiint_M x^1 x^2\, dm,$$

so that with respect to an orthonormal basis $\mathcal{B}$, the inertia tensor is represented by the symmetric matrix

$$I = \begin{pmatrix} I_{11} & I_{12} & I_{13} \\ I_{21} & I_{22} & I_{23} \\ I_{31} & I_{32} & I_{33} \end{pmatrix}.$$

The diagonal components I_{11}, I_{22}, I_{33} are the **moments of inertia** of the rigid body M with respect to the coordinate axes Ox_1, Ox_2, Ox_3, respectively. The off-diagonal components I_{12}, I_{23}, I_{31} are the **polar moments of inertia** or the **products of inertia** of the rigid body M.

Example 6.3

We want to find the inertia tensor of a homogeneous rectangular plate with sides a and b and total mass m, assuming that the rotation preserves the center of mass O. We choose a coordinate system (corresponding to an orthonormal basis) with origin at the center of mass O, with x-axis parallel to the side of length a, y-axis parallel to the side of length b, z-axis perpendicular to the plate (Fig. 6.2), and

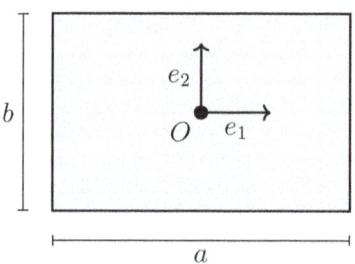

Fig. 6.2 Rectangular plate from Example 6.3

6.1 Inertia Tensor

adjust our previous formulas to double integrals. Since the plate is assumed to be homogeneous, it has a constant *mass density* equal to

$$\rho = \frac{\text{total mass}}{\text{area}} = \frac{m}{ab}.$$

Then

$$\underbrace{I_{11}}_{I_{xx}} = \int_{-\frac{a}{2}}^{\frac{a}{2}} \int_{-\frac{b}{2}}^{\frac{b}{2}} (y^2 + \underbrace{z^2}_{=0}) \underbrace{\rho}_{\frac{m}{ab}} \, dy \, dx$$

$$= \frac{m}{ab} a \int_{-\frac{b}{2}}^{\frac{b}{2}} y^2 \, dy$$

$$= \frac{m}{b} \left[\frac{y^3}{3} \right]_{-\frac{b}{2}}^{\frac{b}{2}} = \frac{m}{12} b^2.$$

Similarly,

$$\underbrace{I_{22}}_{I_{yy}} = \frac{m}{12} a^2,$$

and

$$\underbrace{I_{33}}_{I_{zz}} = \int_{-\frac{a}{2}}^{\frac{a}{2}} \int_{-\frac{b}{2}}^{\frac{b}{2}} (x^2 + y^2) \rho \, dy \, dx = \frac{m}{12} (a^2 + b^2)$$

turns out to be just the sum of I_{11} and I_{22}.
Furthermore,

$$I_{23} = I_{32} = -\int_{-\frac{a}{2}}^{\frac{a}{2}} \int_{-\frac{b}{2}}^{\frac{b}{2}} y \underbrace{z}_{=0} \rho \, dy \, dx = 0,$$

and, similarly, $I_{31} = I_{13} = 0$. Finally, we have

$$I_{21} = I_{21} = -\int_{-\frac{a}{2}}^{\frac{a}{2}} \int_{-\frac{b}{2}}^{\frac{b}{2}} xy \rho \, dy \, dx = -\frac{m}{ab} \underbrace{\left(\int_{-\frac{a}{2}}^{\frac{a}{2}} x \, dx \right)}_{=0} \underbrace{\left(\int_{-\frac{b}{2}}^{\frac{b}{2}} y \, dy \right)}_{=0}.$$

because the integral of an odd function on a symmetric interval is 0

We conclude that the inertia tensor is given by the matrix

$$\frac{m}{12}\begin{pmatrix} b^2 & 0 & 0 \\ 0 & a^2 & 0 \\ 0 & 0 & a^2+b^2 \end{pmatrix}.$$

□

◀

EXERCISE 6.4. Compute the inertia tensor of the same plate, but now with center of rotation O coinciding with a vertex of the rectangular plate.

6.1.3 Moment of Inertia About any Axis

We compute the moment of inertia of the body M about an axis through the point O and defined by the unit vector u (Fig. 6.3).

The **moment of inertia** of an infinitesimal region of M around P with respect to the axis defined by u is

$$dI = \underbrace{r^2}_{\substack{r \text{ is the distance} \\ \text{from } P \text{ to the axis}}} \underbrace{dm}_{\substack{\text{infinitesimal} \\ \text{mass}}} = \|u \times \mathbf{x}\|^2 \, dm,$$

where the last equality follows from the fact that $\|u \times x\| = \|u\| \|x\| \sin\alpha = r$, since u is a unit vector. Hence, the **total moment of inertia** of M with respect to the axis given by u is

$$I_u := \iiint_M \|u \times \mathbf{x}\|^2 \, dm \geq 0.$$

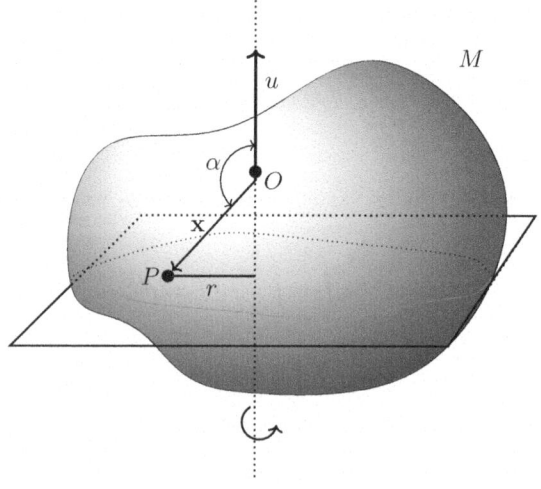

Fig. 6.3 Computing the moment of inertia of a rigid body about an axis

6.1 Inertia Tensor

Fig. 6.4 Rectangular plate from Example 6.5

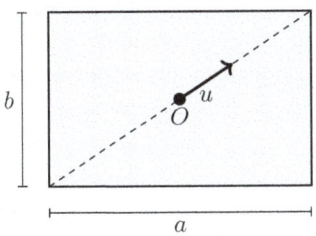

This is very similar to the total kinetic energy E: just replace ω by u and omit the factor $\frac{1}{2}$. By the earlier computations, we conclude that

$$I_u = I_{ij} u^i u^j,$$

where I_{ij} is the inertia tensor. This formula shows that *the total moment of inertia of the rigid body M with respect to an arbitrary axis passing through the point O is determined only by the inertia tensor of the rigid body.*

Example 6.5

For the rectangular plate in Example 6.3, we now want to compute the moment of inertia with respect to the diagonal of the plate (Fig. 6.4).
We choose the unit vector $u = \frac{1}{\sqrt{a^2+b^2}}(ae_1 + be_2)$ (the other possibility is the negative of this vector, yielding the same result), so that

$$u^1 = \frac{a}{\sqrt{a^2+b^2}}, \quad u^2 = \frac{b}{\sqrt{a^2+b^2}}, \quad u^3 = 0$$

and use the matrix for I found in Example 6.3. The moment of inertia is

$$\begin{aligned}
I_u &= I_{ij} u^i u^j \\
&= \begin{pmatrix} \frac{a}{\sqrt{a^2+b^2}} & \frac{b}{\sqrt{a^2+b^2}} & 0 \end{pmatrix} \begin{pmatrix} \frac{m}{12}b^2 & 0 & 0 \\ 0 & \frac{m}{12}a^2 & 0 \\ 0 & 0 & \frac{m}{12}(a^2+b^2) \end{pmatrix} \begin{pmatrix} \frac{a}{\sqrt{a^2+b^2}} \\ \frac{b}{\sqrt{a^2+b^2}} \\ 0 \end{pmatrix} \\
&= \frac{m}{6} \frac{a^2 b^2}{a^2+b^2}.
\end{aligned}$$

◻

◂

EXERCISE 6.6. Double-check the above result for the moment of inertia of the rectangular plate in Example 6.3 with respect to the diagonal of the plate, now by using the inertia tensor computed in Exercise 6.4 (with center of rotation O in a vertex belonging also to that diagonal).

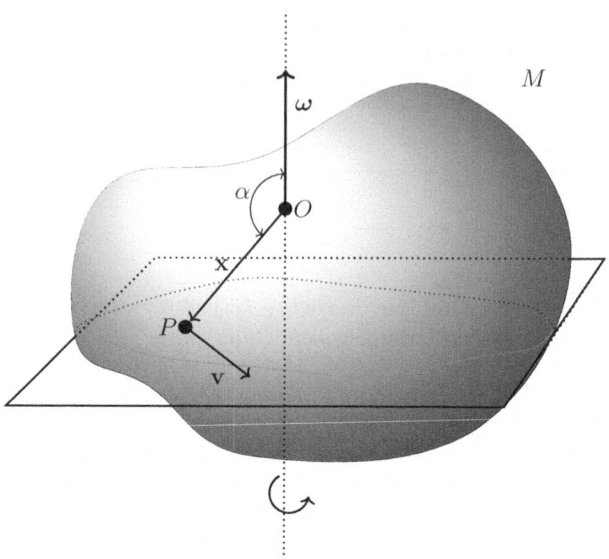

Fig. 6.5 Computing the angular momentum of a rigid body, M, rotating about an axis

EXERCISE 6.7. Compute the moment of inertia of the rectangular plate in Example 6.3 with respect to:

(1) an axis perpendicular to the plate and passing through its center of mass, and
(2) an axis perpendicular to the plate and passing through one vertex.

6.1.4 Angular Momentum

Let M be a body rotating with angular velocity $\boldsymbol{\omega}$ about an axis through the point O. Let $\mathbf{x} = \overrightarrow{OP}$ be the position vector of a point P and $\mathbf{v} = \boldsymbol{\omega} \times \mathbf{x}$ the linear velocity of P (Fig. 6.5).

Then the **angular momentum** of an infinitesimal region of M around P is

$$dL = (\mathbf{x} \times \mathbf{v})\, dm,$$

6.1 Inertia Tensor

so that the **total angular momentum**[4] of M is

$$L = \iiint_M (\mathbf{x} \times (\boldsymbol{\omega} \times \mathbf{x})) \, dm \, .$$

We need to work out $\mathbf{x} \times (\boldsymbol{\omega} \times \mathbf{x})$ for vectors $\mathbf{x}$ and $\boldsymbol{\omega}$ in 3-dimensional space. We apply the following identity[5] for the triple vector product:

$$\mathbf{x} \times (\boldsymbol{\omega} \times \mathbf{x}) = (\mathbf{x} \cdot \mathbf{x})\boldsymbol{\omega} - (\boldsymbol{\omega} \cdot \mathbf{x})\mathbf{x} \, . \tag{6.2}$$

Let $\mathcal{B} = \{e_1, e_2, e_3\}$ be an orthonormal basis of $\mathbb{R}^3$. Then, replacing the following equalities

$$\boldsymbol{\omega} = \omega^i e_i = \delta^i_j \omega^j e_i \tag{6.3}$$

$$\mathbf{x} = x^i e_i = \delta^i_k x^k e_i \tag{6.4}$$

$$\mathbf{x} \cdot \mathbf{x} = \delta_{k\ell} x^k x^\ell \tag{6.5}$$

$$\boldsymbol{\omega} \cdot \mathbf{x} = \delta_{j\ell} \omega^j x^\ell \tag{6.6}$$

into the identity (6.2), we obtain

$$\mathbf{x} \times (\boldsymbol{\omega} \times \mathbf{x}) = \underbrace{(\delta_{k\ell} x^k x^\ell)}_{(6.5)} \underbrace{\delta^i_j \omega^j e_i}_{(6.3)} - \underbrace{(\delta_{j\ell} \omega^j x^\ell)}_{(6.6)} \underbrace{\delta^i_k x^k e_i}_{(6.4)} = (\delta^i_j \delta_{k\ell} - \delta^i_k \delta_{j\ell}) \omega^j x^k x^\ell e_i \, .$$

Therefore, the total angular momentum is

$$L = L^i e_i \, ,$$

where the components L^i are

$$\boxed{L^i = (\delta^i_j \delta_{k\ell} - \delta^i_k \delta_{j\ell}) \, \omega^j \iiint_M x^k x^\ell \, dm} \, .$$

---

[4] Just like the angular velocity, the angular momentum is not an honest vector, but only a *pseudovector*, since there is an issue with orientation. In this subsection, we should thus assume that we work with an *oriented* orthonormal basis $\{e_1, e_2, e_3\}$ of $\mathbb{R}^3$, so that $e_1 \times e_2 = e_3$ (and not $-e_3$). This amounts to assuming that the change of basis matrix L from the standard basis has $\det L = 1$ (and not -1).

[5] To prove this vector equality use coordinates, consider only the case in which $\boldsymbol{\omega}$ is a standard basis vector and then use the linearity in $\boldsymbol{\omega}$.

The above expression for L^i can be written in terms of the inertia tensor I_{ij} as

$$L^i = \delta^{ik} I_{kj} \omega^j ,$$

which corresponds to the matrix form $L = I\omega$. We see that the angular momentum L is proportional (or parallel) to the angular velocity ω only when ω is an eigenvector of the inertia tensor I.

Example 6.8

Suppose the rectangular plate in the previous examples is rotating about an axis through the center of mass O with angular velocity

$$\omega = e_1 + 2e_2 + 3e_3, \quad \text{or} \quad \begin{pmatrix} \omega^1 \\ \omega^2 \\ \omega^3 \end{pmatrix} = \begin{pmatrix} 1 \\ 2 \\ 3 \end{pmatrix}.$$

We want to compute its angular momentum.
The inertia tensor is given by the matrix I_{ij} found in Example 6.3:

$$\begin{pmatrix} \frac{m}{12}b^2 & 0 & 0 \\ 0 & \frac{m}{12}a^2 & 0 \\ 0 & 0 & \frac{m}{12}(a^2+b^2) \end{pmatrix}.$$

The total angular momentum has components given by

$$\begin{pmatrix} L^1 \\ L^2 \\ L^3 \end{pmatrix} = \begin{pmatrix} \frac{m}{12}b^2 & 0 & 0 \\ 0 & \frac{m}{12}a^2 & 0 \\ 0 & 0 & \frac{m}{12}(a^2+b^2) \end{pmatrix} \begin{pmatrix} 1 \\ 2 \\ 3 \end{pmatrix} = \begin{pmatrix} \frac{m}{12}b^2 \\ \frac{m}{6}a^2 \\ \frac{m}{4}(a^2+b^2) \end{pmatrix},$$

so that

$$L = \frac{m}{12}b^2 e_1 + \frac{m}{6}a^2 e_2 + \frac{m}{4}(a^2+b^2) e_3.$$

◻

◀

6.1.5 Principal Moments of Inertia

Observe that the inertia tensor of a rigid body M is *symmetric* and recall the *Spectral Theorem* (Theorem 4.9). Then we know that an orthonormal eigenbasis $\{\tilde{e}_1, \tilde{e}_2, \tilde{e}_3\}$ exists for the inertia tensor. Let I_1, I_2, I_3 be the corresponding eigenvalues. The matrix representing the inertia tensor with respect to this eigenbasis is

$$\begin{pmatrix} I_1 & 0 & 0 \\ 0 & I_2 & 0 \\ 0 & 0 & I_3 \end{pmatrix}.$$

6.1 Inertia Tensor

The orthonormal eigenbasis gives a preferred coordinate system in which to formulate a problem pertaining to rotation of this body. The axes of the eigenvectors are called the **principal axes of inertia** of the rigid body M. The eigenvalues I_i are called the **principal moments of inertia**.

For instance, if a homogeneous body is symmetric with respect to the xy-plane, then the polar moments of inertia $I_{23} = I_{32}$ and $I_{13} = I_{31}$ vanish, thus the z-axis is necessarily a principal axis (because of the block-form of I).

The *principal moments of inertia* are the moments of inertia with respect to the *principal axes of inertia*, hence they are non-negative

$$I_1, I_2, I_3 \geq 0.$$

A rigid body is called

(1) an **asymmetric top** if $I_1 \neq I_2 \neq I_3 \neq I_1$;
(2) a **symmetric top** if exactly two eigenvalues are equal, say $I_1 = I_2 \neq I_3$: any axis passing through the plane determined by $\tilde{e}_1$ and $\tilde{e}_2$ is then a principal axis of inertia;
(3) a **spherical top** if $I_1 = I_2 = I_3$: any axis passing through O is a principal axis of inertia.

With respect to the eigenbasis $\{\tilde{e}_1, \tilde{e}_2, \tilde{e}_3\}$ the *kinetic energy* is

$$E = \frac{1}{2}\left(I_1\tilde{\omega}^1\tilde{\omega}^1 + I_2\tilde{\omega}^2\tilde{\omega}^2 + I_3\tilde{\omega}^3\tilde{\omega}^3\right),$$

where $\omega = \tilde{\omega}^i \tilde{e}_i$, with $\tilde{\omega}^i$ the components of the angular velocity with respect to the basis $\{\tilde{e}_1, \tilde{e}_2, \tilde{e}_3\}$. In particular, we see that the kinetic energy can be conserved, even if the angular velocity ω changes, as long as the above combination of squares is preserved. This is related to the phenomenon of *precession*.

The surface determined by the equation (with respect to the coordinates x, y, z)

$$I_1 x^2 + I_2 y^2 + I_3 z^2 = 1 \tag{6.7}$$

is called the **ellipsoid of inertia**. The symmetry axes of the ellipsoid coincide with the principal axes of inertia. Note that for a spherical top, the ellipsoid of inertia is actually a sphere.

The ellipsoid of inertia gives the moment of inertia with respect to any axis as follows: Consider an axis given by the unit vector u and let $p = cu$ be a vector of intersection of the axis with the ellipsoid of inertia, where c is ($\pm$) the distance to O of the intersection of the axis with the ellipsoid of inertia. The moment of inertia with respect to this axis is (Fig. 6.6)

$$I = I_{ij} u^i u^j = \frac{1}{c^2} I_{ij} p^i p^j = \frac{1}{c^2},$$

where the last equality follows from the fact that, since p is on the ellipsoid, then $I_{ij} u^i u^j = 1$ by Eq. 6.7.

Fig. 6.6 Ellipsoid of inertia

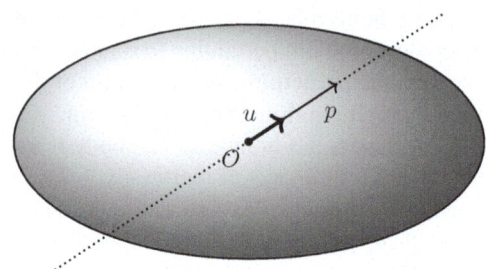

Example 6.9

The principal axes of inertia for the rectangular plate in Example 6.3 are the axes parallel to the sides and the axis perpendicular to the plate. The corresponding principal moments of inertia are

$$I_{11} = \frac{m}{12}b^2, \qquad I_{22} = \frac{m}{12}a^2 \quad \text{and} \quad I_{33} = \frac{m}{12}(a^2 + b^2).$$

If $a = b$, that is, if the rectangle is a square, we have a *symmetric top*. □
◂

EXERCISE 6.10. The following systems are regarded as *rigid*, i.e., as systems where the distances between particles remain constant, and where the origin is at the center of mass. For each system, choose an orthonormal basis and determine the inertia tensor and the ellipsoid of inertia, giving in each case the principal axes of inertia and the principal moments of inertia.

(1) A molecule consisting of n atoms with masses m_i, $i = 1, \ldots, n$, all along a straight line L with relative distance ℓ_{ij} between atom i and atom j.
(2) A molecule made up of three atoms, which lie in the vertices of an isosceles triangle ABC with basis length $BC = a$ and height h. The atoms in positions B und C have mass m_1, the atom in position A has mass m_2.
(3) A molecule composed of four atoms, all with the same mass m, lying in the vertices of a regular tetrahedron with edges of length a.

EXERCISE 6.11. Let K be a homogeneous rigid body of mass m and shaped like a parallelepiped with orthogonal sides a, b, c. The body K rotates around one of its diagonals (going necessarily through its center of mass) with angular velocity ω.

(1) Find the principal axes and principal moments of inertia of K.
(2) Find the equation of its ellipsoid of inertia.
(3) Find the kinetic energy of K.
(4) Find the angular momentum of K.

6.2 Stress Tensor

6.2.1 Physical Preliminaries

Let us consider a rigid body M acted upon by external forces but in *static equilibrium*, and let us consider an infinitesimal region dM around a point P. There are two types of external forces:

(1) The **body forces**, that is forces whose magnitude is proportional to the volume/mass of the region. For instance, *gravity*, *attractive force* or the *centrifugal force*.
(2) The **surface forces**, that is forces exerted on the surface of the element by the material surrounding it. These are forces whose magnitude is proportional to the area of the region in consideration.

The *surface force per unit area* is called the **stress**. We will concentrate on **homogeneous stress**, that is stress that does not depend on the location of the element in the body, but depends only on the orientation of the surface given by its tangent plane. Moreover, we assume that the body in consideration is in *static equilibrium*.

Remark 6.12 It was the concept of *stress* in mechanics that originally led to the invention of tensors, also etymologically. The English word *stress* relates to tension, leading to the choice of *tensor* or, in French, *tenseur*.

Choose an orthonormal basis $\{e_1, e_2, e_3\}$ and the plane Π_1 through P parallel to the $e_2 e_3$ coordinate plane. The vector e_1 is normal to this plane. Let ΔA_1 be the area of the slice of the infinitesimal region around P cut by the plane and let ΔF be the force acting on that slice. In this orthonormal set-up (cf. Sect. 4.3.3), we write ΔF as a vector in terms of its components

$$\Delta F = \Delta F^1 e_1 + \Delta F^2 e_e + \Delta F^3 e_3$$

and, since the stress is the surface force per unit area, we define

$$\sigma^{1j} := \lim_{\Delta A_1 \to 0} \frac{\Delta F^j}{\Delta A_1}, \quad \text{for } j = 1, 2, 3.$$

Similarly, we can consider planes parallel to the other coordinate planes and define (Fig. 6.7)

$$\boxed{\sigma^{ij} := \lim_{\Delta A_i \to 0} \frac{\Delta F^j}{\Delta A_i}}.$$

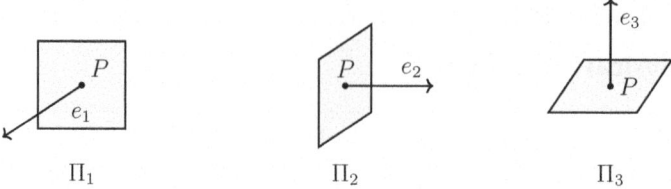

Fig. 6.7 Planes through the point P and parallel to the coordinate planes, for the computation of the stress tensor

Fig. 6.8 Computing the stress tensor across a planar slice through the point P

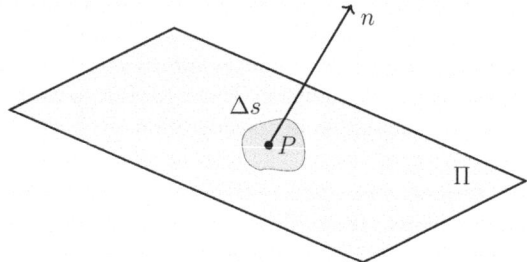

It turns out that the resulting nine numbers σ^{ij} are the components of a contravariant 2-tensor called the **stress tensor**. As we are restricting to orthonormal bases, the type of this tensor gets blurred; cf. Sect. 4.3.3. The stress tensor encodes the mechanical stresses on an object.

We now compute the stress across *other* slices through P, that is, across other planes with other normal vectors. Let Π be a plane passing through P, $\mathbf{n}$ a unit vector through P perpendicular to the plane Π, $\Delta s = \Pi \cap dM$ the area of a small element of the plane Π containing P and ΔF the force acting on that element (Fig. 6.8).

CLAIM 6.13 The stress at P across the surface perpendicular to $\mathbf{n}$ is

$$\sigma(\mathbf{n}) := \lim_{\Delta s \to 0} \frac{\Delta F}{\Delta s} = \sigma^{ij}(\mathbf{n} \cdot e_i)e_j.$$

It follows from the claim that the stress σ is a vector-valued function that depends linearly on the normal $\mathbf{n}$ to the surface element.

Proof Consider the tetrahedron $OA_1A_2A_3$ bound by the *triangular slice* on the plane Π having area Δs and three triangles on planes parallel to the coordinate planes (Fig. 6.9)

Consider all external forces acting on this tetrahedron, which we regard as a volume element of the rigid body:

6.2 Stress Tensor

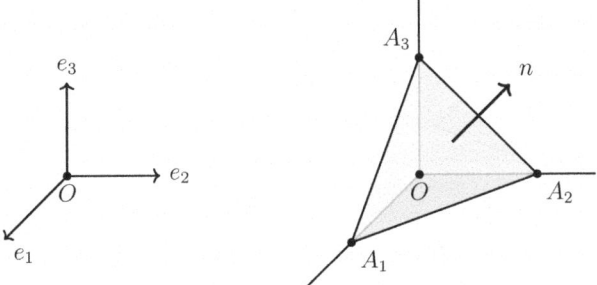

Fig. 6.9 Tetrahedron for the proof of Claim 6.13

(1) *Body forces* amounting to $f \cdot \Delta v$, where f is the force per unit of volume and Δv is the volume of the tetrahedron. We actually do not know these forces, but we will see later that these are not relevant.
(2) *Surface forces* amounting to the sum of the forces on each of the four sides of the tetrahedron.

We want to assess each of the four surface contributions due to the surface forces. If Δs is the area of the slice on the plane Π, the contribution of that slice is, by definition of stress equal to

$$\sigma(\mathbf{n})\Delta s.$$

If Δs_1 is the area of the slice on the plane with normal $-e_1$, the contribution of that slice is

$$-\sigma^{1j} e_j \Delta s_1,$$

and, similarly, the contributions of the other two slices are

$$-\sigma^{2j} e_j \Delta s_2 \quad \text{and} \quad -\sigma^{3j} e_j \Delta s_3.$$

Note that the minus sign comes from the fact that we use everywhere outside pointing normals (Fig. 6.10).

So the total surface force is

$$\sigma(\mathbf{n})\Delta s - \sigma^{1j} e_j \Delta s_1 - \sigma^{2j} e_j \Delta s_2 - \sigma^{3j} e_j \Delta s_3.$$

Since there is static equilibrium, the sum of all (body and surface) forces must be zero

$$f \Delta v + \sigma(\mathbf{n})\Delta s - \sigma^{ij} e_j \Delta s_i = 0.$$

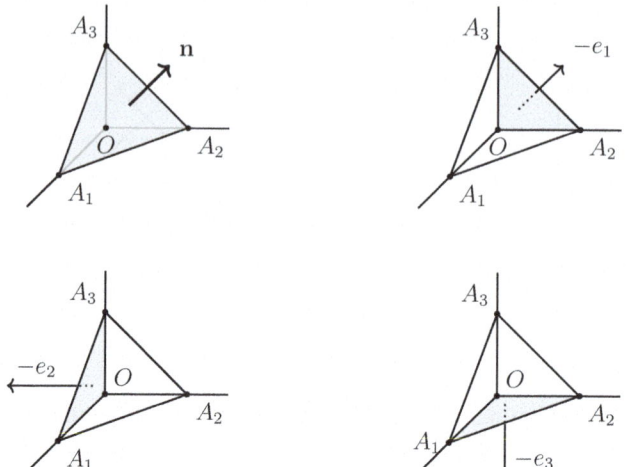

Fig. 6.10 The four faces of the tetrahedron $OA_1A_2A_3$ and their corresponding outward-pointing unit normal vectors

Fig. 6.11 The convex angle between vectors e_i and $\mathbf{n}$

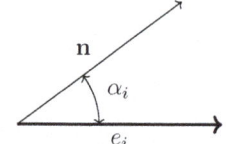

The term $f\Delta v$ can be neglected when Δs is small, as it contains terms of higher order (in fact, $\Delta v \to 0$ faster than $\Delta s \to 0$). We conclude that

$$\sigma(\mathbf{n})\Delta s = \sigma^{ij} e_j \Delta s_i \,.$$

It remains to relate Δs to $\Delta s_1, \Delta s_2, \Delta s_3$. The side with area Δs_i is the orthogonal projection of the side with area Δs onto the plane with normal e_i. The scaling factor for the area under projection is $\cos \alpha_i$, where α_i is the convex angle between the plane normal vectors (Fig. 6.11)

$$\frac{\Delta s_i}{\Delta s} = \cos \alpha_i = \cos \alpha_i \|\mathbf{n}\| \|e_i\| = \mathbf{n} \cdot e_i \,.$$

Therefore,

$$\sigma(\mathbf{n})\Delta s = \sigma^{ij} e_j (\mathbf{n} \cdot e_i) \Delta s$$

or, equivalently,

$$\sigma(\mathbf{n}) = \sigma^{ij} (\mathbf{n} \cdot e_i) e_j \,.$$

6.2 Stress Tensor

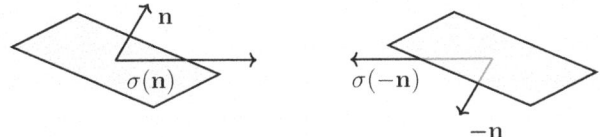

Fig. 6.12 Stress changes sign under orientation flipping of the normal to the plane

Note that, in the above formula, the quantities $\mathbf{n} \cdot e_i$ are the coordinates of $\mathbf{n}$ with respect to the orthonormal basis $\{e_1, e_2, e_3\}$, namely

$$\mathbf{n} = (\mathbf{n} \cdot e_1)e_1 + (\mathbf{n} \cdot e_2)e_2 + (\mathbf{n} \cdot e_3)e_3 = n^1 e_1 + n^2 e_2 + n^3 e_3.$$

$\square$

Remark 6.14 For homogeneous stress, the stress tensor σ^{ij} does not depend on the point P. However, when we flip the orientation of the normal to the plane, the stress tensor changes sign. In other words, if $\sigma(\mathbf{n})$ is the stress across a surface with normal $\mathbf{n}$, then

$$\sigma(-\mathbf{n}) = -\sigma(\mathbf{n}).$$

The stress considers orientation as if the forces on each side of the surface have to balance each other in static equilibrium (Fig. 6.12).

$\square$

6.2.2 Principal Stresses

The following claim is also known as the Boltzmann Axiom.

CLAIM 6.15 The stress tensor is a *symmetric tensor*, that is $\sigma^{ij} = \sigma^{ji}$.

Proof Consider an infinitesimal cube of side $\Delta \ell$ surrounding P and with faces parallel to the coordinate planes (Fig. 6.13).

The force acting on each of the six faces of the cube are:

- $\sigma^{1j} \Delta A_1 e_j$ and $-\sigma^{1j} \Delta A_1 e_j$, respectively for the front and the back faces, $ABB'A'$ and $DCC'D'$;
- $\sigma^{2j} \Delta A_2 e_j$ and $-\sigma^{2j} \Delta A_2 e_j$, respectively for the right and the left faces $BCC'B'$ and $ADD'A'$;
- $\sigma^{3j} \Delta A_3 e_j$ and $-\sigma^{3j} \Delta A_3 e_j$, respectively for the top and the bottom faces $ABCD$ and $A'B'C'D'$,

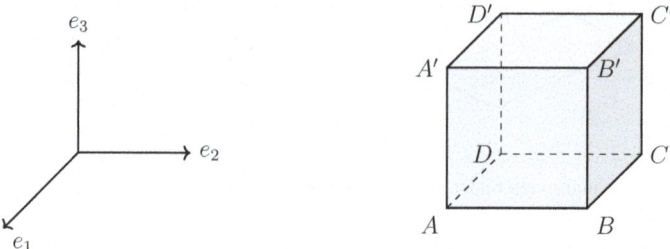

Fig. 6.13 Cube with faces parallel to the coordinate planes for the proof of Claim 6.15

where $\Delta A_1 = \Delta A_2 = \Delta A_3 = \Delta s = (\Delta \ell)^2$ is the common face area. We compute now the *torque* τ, assuming the forces are applied at the center of each face, whose distance to the center point P is $\frac{1}{2}\Delta\ell$. Recall that the **torque** (or *moment of force*) is the tendency of a force to twist or rotate an object. It is given by the cross product of the distance vector and the force vector.

$$\begin{aligned}\tau = & \tfrac{\Delta\ell}{2}e_1 \times \sigma^{1j}\Delta s\, e_j + \left(-\tfrac{\Delta\ell}{2}e_1\right) \times (-\sigma^{1j}\Delta s\, e_j) \\ & + \tfrac{\Delta\ell}{2}e_2 \times \sigma^{2j}\Delta s\, e_j + \left(-\tfrac{\Delta\ell}{2}e_2\right) \times (-\sigma^{2j}\Delta s\, e_j) \\ & + \tfrac{\Delta\ell}{2}e_2 \times \sigma^{3j}\Delta s\, e_j + \left(-\tfrac{\Delta\ell}{2}e_3\right) \times (-\sigma^{3j}\Delta s\, e_j) \\ = & \Delta\ell \Delta s\, (e_i \times \sigma^{ij} e_j) \\ = & \Delta\ell \Delta s \left((\sigma^{23} - \sigma^{32})e_1 + (\sigma^{31} - \sigma^{13})e_2 + (\sigma^{12} - \sigma^{21})e_3\right).\end{aligned}$$

Since the equilibrium is static, then $\tau = 0$, so that $\sigma^{ij} = \sigma^{ji}$. □

We can hence write

$$\sigma = \begin{pmatrix} \sigma^{11} & \sigma^{12} & \sigma^{13} \\ \sigma^{12} & \sigma^{22} & \sigma^{23} \\ \sigma^{13} & \sigma^{23} & \sigma^{33} \end{pmatrix},$$

where the diagonal entries σ^{11}, σ^{22} and σ^{33} are the **normal components**, that is the components of the forces perpendicular to the coordinate planes and the remaining entries σ^{12}, σ^{13} and σ^{23} are the **shear components**, that is the components of the forces parallel to the coordinate planes.

Since the stress tensor is symmetric, again by the *Spectral Theorem* (Theorem 4.9), it can be orthogonally diagonalized. That is, with respect to an orthonormal eigenbasis, it is given by

$$\begin{pmatrix} \sigma^1 & 0 & 0 \\ 0 & \sigma^2 & 0 \\ 0 & 0 & \sigma^3 \end{pmatrix},$$

6.2 Stress Tensor

where now σ^1, σ^2 and σ^3 are the **principal stresses**, that is the *eigenvalues of* σ. The eigenspaces of σ are the **principal directions** and the shear components disappear for the **principal planes**, i.e., the planes orthogonal to the principal directions.

6.2.3 Special Forms of the Stress Tensor

We consider the stress tensor with respect to an orthonormal eigenbasis or another special basis, so that the corresponding matrix has a simpler form. We use the following terminology:

- **Uniaxial stress** for a stress tensor given by

$$\begin{pmatrix} \sigma & 0 & 0 \\ 0 & 0 & 0 \\ 0 & 0 & 0 \end{pmatrix}$$

Example 6.16

This is the stress tensor in a long vertical rod loaded by hanging a weight on the end. ◻

◀

- **Plane stress** or **biaxial stress** for a stress tensor given by

$$\begin{pmatrix} \sigma^1 & 0 & 0 \\ 0 & \sigma^2 & 0 \\ 0 & 0 & 0 \end{pmatrix}$$

Example 6.17

This is the stress tensor in a plate on which forces are applied parallel to the plate (Fig. 6.14). ◻

◀

Fig. 6.14 Forces originating plane stress

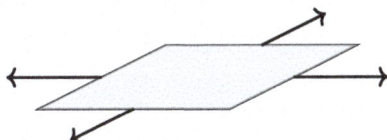

Fig. 6.15 Forces originating pure shear

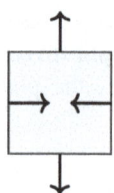 

- **Pure shear** for a stress tensor given by

$$\begin{pmatrix} -\sigma & 0 & 0 \\ 0 & \sigma & 0 \\ 0 & 0 & 0 \end{pmatrix} \quad \text{or} \quad \begin{pmatrix} 0 & \sigma & 0 \\ \sigma & 0 & 0 \\ 0 & 0 & 0 \end{pmatrix}. \tag{6.8}$$

This is special case of the biaxial stress, in the case where $\sigma^1 = -\sigma^2$. The first matrix in (6.8) represents the stress tensor written with respect to an eigenbasis, whereas the second matrix represents the stress tensor written with respect to an orthonormal basis obtained by rotating an eigenbasis by 45° about the third axis. The relation between the two matrix representations is given by

$$\begin{pmatrix} 0 & \sigma & 0 \\ \sigma & 0 & 0 \\ 0 & 0 & 0 \end{pmatrix} = \underbrace{\begin{pmatrix} \frac{\sqrt{2}}{2} & \frac{\sqrt{2}}{2} & 0 \\ -\frac{\sqrt{2}}{2} & \frac{\sqrt{2}}{2} & 0 \\ 0 & 0 & 1 \end{pmatrix}}_{{}^t L} \begin{pmatrix} -\sigma & 0 & 0 \\ 0 & \sigma & 0 \\ 0 & 0 & 0 \end{pmatrix} \underbrace{\begin{pmatrix} \frac{\sqrt{2}}{2} & -\frac{\sqrt{2}}{2} & 0 \\ \frac{\sqrt{2}}{2} & \frac{\sqrt{2}}{2} & 0 \\ 0 & 0 & 1 \end{pmatrix}}_{L}$$

where L is the matrix of the change of coordinates (Fig. 6.15).

- **Shear deformation** for a stress tensor given by

$$\begin{pmatrix} 0 & \sigma^{12} & \sigma^{13} \\ \sigma^{12} & 0 & \sigma^{23} \\ \sigma^{13} & \sigma^{23} & 0 \end{pmatrix}$$

(with respect to some orthonormal basis).

Example 6.18

The stress tensor

$$\begin{pmatrix} 2 & -4 & 0 \\ -4 & 0 & 4 \\ 0 & 4 & -2 \end{pmatrix}$$

6.2 Stress Tensor

represents a shear deformation. In fact, one can check that

$$\underbrace{\begin{pmatrix} \frac{\sqrt{2}}{2} & 0 & \frac{\sqrt{2}}{2} \\ 0 & 1 & 0 \\ -\frac{\sqrt{2}}{2} & 0 & \frac{\sqrt{2}}{2} \end{pmatrix}}_{{}^tL} \begin{pmatrix} 2 & -4 & 0 \\ -4 & 0 & 4 \\ 0 & 4 & -2 \end{pmatrix} \underbrace{\begin{pmatrix} \frac{\sqrt{2}}{2} & 0 & -\frac{\sqrt{2}}{2} \\ 0 & 1 & 0 \\ \frac{\sqrt{2}}{2} & 0 & \frac{\sqrt{2}}{2} \end{pmatrix}}_{L} = \begin{pmatrix} 0 & 0 & -2 \\ 0 & 0 & 4\sqrt{2} \\ -2 & 4\sqrt{2} & 0 \end{pmatrix}$$

□

◂

- **Hydrostatic pressure** with stress tensor given by

$$\begin{pmatrix} -p & 0 & 0 \\ 0 & -p & 0 \\ 0 & 0 & -p \end{pmatrix},$$

where $p \neq 0$ is the pressure. Here all eigenvalues are equal to $-p$.

Example 6.19

Pressure of a fluid on a bubble. □

◂

6.2.4 Stress Invariants

Let A be a 3×3 matrix with entries a^{ij}. The *characteristic polynomial* $p_A(\lambda)$ of A is invariant under conjugation (see Sect. 2.4.2), so its coefficients remain unchanged.

$$p_A(\lambda) = \det(A - \lambda I) = \det \begin{pmatrix} a^{11} - \lambda & a^{12} & a^{13} \\ a^{21} & a^{22} - \lambda & a^{23} \\ a^{31} & a^{32} & a^{33} - \lambda \end{pmatrix}$$

$$= -\lambda^3 + \operatorname{tr} A \lambda^2$$
$$- \underbrace{\left(a^{11}a^{22} + a^{11}a^{33} + a^{22}a^{33} - a^{12}a^{21} - a^{23}a^{32} - a^{13}a^{31}\right)}_{\text{quadratic expression in the entries of } A} \lambda + \det A.$$

Applying this to the stress tensor $\sigma = A$, we obtain some **stress invariants**, namely:

$$I_1 := \operatorname{tr} \sigma = \sigma^{11} + \sigma^{22} + \sigma^{33}$$
$$I_2 := \sigma^{12}\sigma^{12} + \sigma^{23}\sigma^{23} + \sigma^{13}\sigma^{13} - \sigma^{11}\sigma^{22} - \sigma^{22}\sigma^{33} - \sigma^{33}\sigma^{11}$$
$$I_3 := \det \sigma.$$

That means, that the above quantities I_1, I_2 and I_3 are invariant when we change the orthonormal basis. Indeed, by contravariance, when we change basis via a matrix L, the matrix of the stress tensor changes from σ to $\tilde{\sigma} = {}^t\Lambda \sigma \Lambda$, where $\Lambda = L^{-1}$. But since we are restricting to orthonormal bases, we have that the change of basis matrix is orthogonal, i.e., $\Lambda = {}^t L$, so this is in fact a conjugation: $\tilde{\sigma} = L \sigma L^{-1}$.

EXERCISE 6.20. Let $\mathcal{E} := \{e_1, e_2, e_3\}$ be an orthonormal basis of $\mathbb{R}^3$ (with the standard inner product). Let σ be a stress tensor given with respect to the basis $\mathcal{E}$ by

$$\sigma = [\sigma^{ij}] := \begin{bmatrix} 0 & 0 & 6 \\ 0 & 1 & 0 \\ 6 & 0 & 5 \end{bmatrix}.$$

(1) Find the principal stresses $\sigma^1, \sigma^2, \sigma^3$ of σ.
(2) Find the principal directions of σ.
(3) Find an orthonormal basis $\mathcal{B} := \{b_1, b_2, b_3\}$ of $\mathbb{R}^3$, with respect to which σ is given by a diagonal matrix.
(4) Find the three stress invariants I_1, I_2 and I_3 of σ.

6.2.5 Decomposition of the Stress Tensor

Any stress tensor σ can be expressed as the sum of two other stress tensors:

- The **hydrostatic stress tensor**

$$\pi \delta^{ij} = \begin{pmatrix} \pi & 0 & 0 \\ 0 & \pi & 0 \\ 0 & 0 & \pi \end{pmatrix},$$

where $\pi := \operatorname{tr}\sigma/3 = (\sigma^{11} + \sigma^{22} + \sigma^{33})/3$. This relates to *volume change*.
- The **deviatoric stress tensor**

$$s^{ij} := \sigma^{ij} - \pi \delta^{ij} = \begin{pmatrix} \sigma^{11} - \pi & \sigma^{12} & \sigma^{13} \\ \sigma^{12} & \sigma^{22} - \pi & \sigma^{23} \\ \sigma^{13} & \sigma^{23} & \sigma^{33} - \pi \end{pmatrix}.$$

This relates to *shape change*.

The **hydrostatic pressure** p is generally defined as the negative one third of the first stress invariant $I_1 = \operatorname{tr}\sigma$, i.e., $p = -\pi$.

Clearly, we have

$$\sigma^{ij} = s^{ij} + \pi \delta^{ij}$$

6.2 Stress Tensor

and, hence, any stress tensor is a sum of a deviatoric stress and a hydrostatic pressure.

Moreover, a shear deformation (see Sect. 6.2.3) is traceless, hence a deviatoric stress.

Actually, the converse is also true:

FACT 6.21 Any deviatoric stress, i.e., any traceless stress, may be represented with respect to some orthonormal basis as a shear deformation, i.e., as a stress tensor of the form

$$\begin{pmatrix} 0 & \tilde{\sigma}^{12} & \tilde{\sigma}^{13} \\ \tilde{\sigma}^{12} & 0 & \tilde{\sigma}^{23} \\ \tilde{\sigma}^{13} & \tilde{\sigma}^{23} & 0 \end{pmatrix}.$$

This important fact follows from the Spectral Theorem (Theorem 4.9).

We conclude that, any stress tensor can be decomposed as the sum of a hydrostatic pressure and a shear deformation.

EXERCISE 6.22. Let $\mathcal{E} := \{e_1, e_2, e_3\}$ be an orthonormal basis of $\mathbb{R}^3$ (with the standard inner product). We consider a stress tensor σ given by the following matrix with respect to the basis $\mathcal{E}$:

$$\sigma = [\sigma^{ij}] := \begin{bmatrix} -2 & 0 & 3 \\ 0 & 2 & 0 \\ 3 & 0 & 2 \end{bmatrix}.$$

(1) Write the above matrix representation of the stress tensor σ as a sum of a deviatoric stress σ_S (that is, σ_S is traceless) and a hydrostatic pressure σ_P (that is, σ_P is a multiple of the identity matrix).
(2) Find an orthonormal basis $\mathcal{D} := \{v_1, v_2, v_3\}$ of $\mathbb{R}^3$, with respect to which the above deviatoric stress σ_S is given by a diagonal matrix, D.
(3) Find an orthonormal basis $\mathcal{B} := \{b_1, b_2, b_3\}$ of $\mathbb{R}^3$, with respect to which the above deviatoric stress σ_S is given by a matrix A with vanishing diagonal elements:

$$A = \begin{bmatrix} 0 & x & y \\ x & 0 & z \\ y & z & 0 \end{bmatrix}$$

for some $x, y, z \in \mathbb{R}$.

Hint: The matrix A must have the same eigenvalues as σ_S, since they are conjugate matrices. Hence, each σ_S and A can be diagonalised to the *same* diagonal matrix D. Moreover, this can be achieved by choosing appropriate *orthonormal* bases in each case, since these matrices are symmetric.

6.3 Strain Tensor

6.3.1 Physical Preliminaries

Consider a slight deformation of a body, where we compare the relative positions of two particles, P and P_1, before and after the deformation:

We have

$$\Delta \tilde{x} = \Delta x + \Delta u,$$

where Δx is the old relative position of P and P_1, $\Delta \tilde{x}$ is their new relative position and Δu is the displacement difference, which hence measures the deformation (Fig. 6.16).

Assume that we have a small homogeneous deformation, that is

$$\Delta u = f(\Delta x),$$

in other words, f is a small linear function independent of the point P. If we write the components of Δu and Δx with respect to an orthonormal basis $\{e_1, e_2, e_3\}$, the function f will be represented by a matrix with entries that we denote by f_{ij},

$$\Delta u_i = f_{ij} \Delta x^j.$$

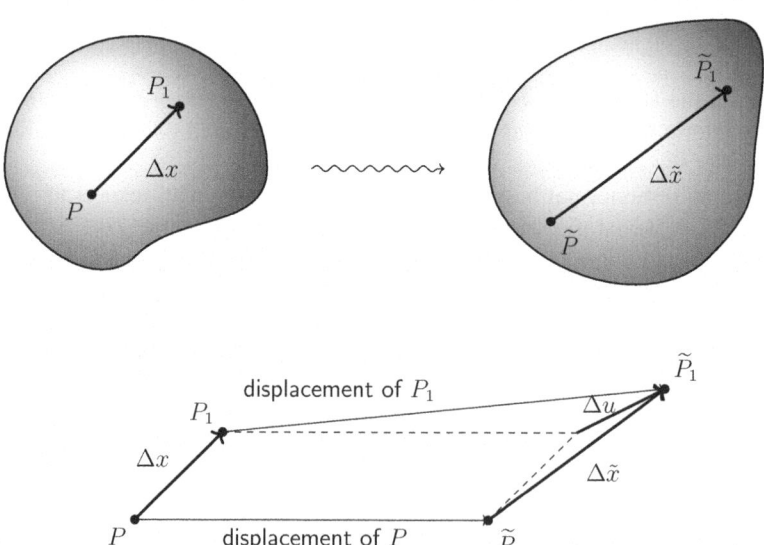

Fig. 6.16 Comparison of relative positions of two particles before and after a slight deformation

6.3 Strain Tensor

The matrix (f_{ij}) can be written as a sum of a symmetric and an antisymmetric matrix as follows:

$$f_{ij} = \varepsilon_{ij} + \omega_{ij},$$

where

$$\varepsilon_{ij} = \frac{1}{2}(f_{ij} + f_{ji})$$

is a symmetric matrix and is called the **strain tensor** or **deformation tensor** and

$$\omega_{ij} = \frac{1}{2}(f_{ij} - f_{ji})$$

is an antisymmetric matrix called the **rotation tensor**. We will next try to understand where these names come from.

Remark 6.23 First we verify that a (small) antisymmetric 3×3 matrix represents a (small) rotation in 3-dimensional space.

FACT 6.24 Let V be a vector space with orthonormal basis $\mathcal{B} = \{e_1, e_2, e_3\}$, and let $\omega = \begin{pmatrix} a \\ b \\ c \end{pmatrix}$. The matrix R_ω of the linear map $V \to V$ defined by $v \mapsto \omega \times v$ with respect to the basis $\mathcal{B}$ is

$$R_\omega = \begin{pmatrix} 0 & -c & b \\ c & 0 & -a \\ -b & a & 0 \end{pmatrix}.$$

Indeed, we have

$$\omega \times v = \begin{pmatrix} a \\ b \\ c \end{pmatrix} \times \begin{pmatrix} x \\ y \\ z \end{pmatrix} = \det \begin{pmatrix} e_1 & e_2 & e_3 \\ a & b & c \\ x & y & z \end{pmatrix}$$

$$= \begin{pmatrix} bz - cy \\ cx - az \\ ay - bx \end{pmatrix} = \begin{pmatrix} 0 & -c & b \\ c & 0 & -a \\ -b & a & 0 \end{pmatrix} \begin{pmatrix} x \\ y \\ z \end{pmatrix}.$$

Note that the matrix $R_\omega = (\omega_{ij}) := \begin{pmatrix} 0 & \omega_{12} & -\omega_{31} \\ -\omega_{12} & 0 & \omega_{23} \\ \omega_{31} & -\omega_{23} & 0 \end{pmatrix}$ corresponds to the cross product with the vector $\omega = \begin{pmatrix} -\omega_{23} \\ -\omega_{31} \\ -\omega_{12} \end{pmatrix}$. □

6.3.2 The Antisymmetric Case: Rotation

Suppose that the matrix (f_{ij}) was already *antisymmetric*, so that

$$\omega_{ij} = f_{ij} \quad \text{and} \quad \varepsilon_{ij} = 0.$$

Note that

$$\omega_{ii} = \frac{1}{2}(f_{ii} - f_{ii}) = 0.$$

By the Fact 6.24, if $\omega = \begin{pmatrix} -\omega_{23} \\ -\omega_{31} \\ -\omega_{12} \end{pmatrix}$, then

$$R_\omega \Delta x = \omega \times \Delta x$$

and the equation

$$\Delta u^i = f_{ij} \Delta x^j \tag{6.9}$$

is equivalent to

$$\Delta u = \omega \times \Delta x,$$

so that

$$\Delta \tilde{x} = \Delta x + \Delta u = \Delta x + \omega \times \Delta x.$$

When ω is small, this represents an infinitesimal rotation of an angle $\|\omega\|$ about the axis $O\omega$ (Fig. 6.17).

In fact, since $\omega \times \Delta x$ is orthogonal to the plane determined by ω and by Δx, it is tangent to the circle with center along the axis $O\omega$ and radius r determined by Δx. Moreover,

$$\|\Delta u\| = \|\omega \times \Delta x\| = \|\omega\| \underbrace{\|\Delta x\| \sin \alpha}_{r},$$

Fig. 6.17 Infinitesimal rotation about the axis $O\omega$

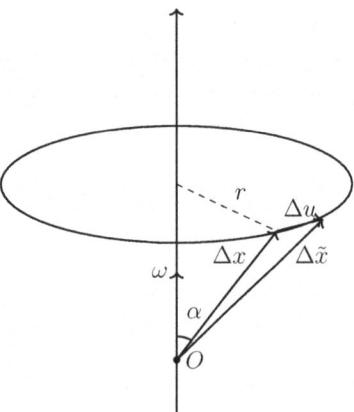

and hence, since the length of an arc of a circle of radius r corresponding to an angle θ is $r\theta$, infinitesimally this represents a rotation by an angle $\|\omega\|$.

6.3.3 The Symmetric Case: Strain

The opposite extreme case is when the matrix f_{ij} was already *symmetric*, so that

$$\varepsilon_{ij} = f_{ij} \quad \text{and} \quad \omega_{ij} = 0.$$

We will see that it is ε_{ij} that encodes the changes in the distances: in fact,

$$\begin{aligned} \|\Delta \tilde{x}\|^2 &= \Delta \tilde{x} \cdot \Delta \tilde{x} = (\Delta x + \Delta u) \cdot (\Delta x + \Delta u) \\ &= \Delta x \cdot \Delta x + 2 \Delta x \cdot \Delta u + \Delta u \cdot \Delta u \\ &\simeq \|\Delta x\|^2 + 2\varepsilon_{ij} \Delta x^i \Delta x^j, \end{aligned} \quad (6.10)$$

where in the last step we neglected the term $\|\Delta u\|^2$ since it is small compared to Δu when $\Delta u \to 0$ and used Eq. 6.9.

Remark 6.25 Even when f_{ij} is not purely symmetric, only the symmetric part ε_{ij} is relevant for the distortion of the distances. In fact, since ω_{ij} is antisymmetric, the term $2\omega_{ij} \Delta x^i \Delta x^j = 0$, so that

$$\|\Delta \tilde{x}\|^2 \simeq \|\Delta x\|^2 + 2 f_{ij} \Delta x^i \Delta x^j = \|\Delta x\|^2 + 2\varepsilon_{ij} \Delta x^i \Delta x^j.$$

□

Recall that a metric tensor (or inner product) encodes the distances among points. It follows that a deformation changes the metric tensor. Let us denote by g the metric before the deformation and by $\tilde{g}$ the metric after the deformation. By definition, we have

$$\|\Delta \tilde{x}\|^2 \overset{\text{def}}{=} \tilde{g}(\Delta \tilde{x}, \Delta \tilde{x}) = \tilde{g}_{ij} \Delta \tilde{x}^i \Delta \tilde{x}^j = \tilde{g}_{ij}(\Delta x^i + \Delta u^i)(\Delta x^j + \Delta u^j) \qquad (6.11)$$

and

$$\|\Delta x\|^2 \overset{\text{def}}{=} g(\Delta x, \Delta x) = g_{ij} \Delta x^i \Delta x^j . \qquad (6.12)$$

For infinitesimal deformations (that is, if $\Delta u \simeq 0$), Eq. 6.11 becomes

$$\|\Delta \tilde{x}\|^2 \simeq \tilde{g}_{ij} \Delta x^i \Delta x^j .$$

This, together with Eqs. 6.10 and 6.12, leads to

$$\tilde{g}_{ij} \Delta x^i \Delta x^j \simeq g_{ij} \Delta x^i \Delta x^j + 2\varepsilon_{ij} \Delta x^i \Delta x^j$$

and hence

$$\varepsilon_{ij} \simeq \frac{1}{2}(\tilde{g}_{ij} - g_{ij}),$$

that is, ε_{ij} measures the change in the metric.

By definition, the strain tensor ε_{ij} is symmetric

$$\mathcal{E} = \begin{pmatrix} \varepsilon_{11} & \varepsilon_{12} & \varepsilon_{13} \\ \varepsilon_{12} & \varepsilon_{22} & \varepsilon_{23} \\ \varepsilon_{13} & \varepsilon_{23} & \varepsilon_{33} \end{pmatrix},$$

where the terms on the diagonal (in green) determine the elongation or the contraction of the body along the coordinate directions e_1, e_2, e_3, and the terms above the diagonal (in orange) are the **shear components** of the strain tensor; that is ε_{ij} is the movement of a line element parallel to Oe_j towards Oe_i. Since it is a symmetric tensor, it can be orthogonally diagonalized (cf. Theorem 4.9), so we can find an orthonormal basis with respect to which $\mathcal{E}$ is given by

$$\begin{pmatrix} \epsilon_1 & 0 & 0 \\ 0 & \epsilon_2 & 0 \\ 0 & 0 & \epsilon_3 \end{pmatrix},$$

The eigenvalues of $\mathcal{E}$ are the **principal coefficients** of the deformation and the eigenspaces are the **principal directions** of the deformation.

6.3.4 Special Forms of the Strain Tensor

We use the following terminology:

(1) **Shear deformation** when $\mathcal{E}$ is traceless,

$$\operatorname{tr}\mathcal{E} = \varepsilon_{11} + \varepsilon_{22} + \varepsilon_{33} = 0.$$

(2) **Uniform compression** when the principal coefficients of $\mathcal{E}$ are equal,

$$\begin{pmatrix} k & 0 & 0 \\ 0 & k & 0 \\ 0 & 0 & k \end{pmatrix}.$$

EXERCISE 6.26. Show that any strain tensor can be written as the sum of a uniform compression and a shear deformation. *Hint:* See Sect. 6.2.5.

6.4 Elasticity Tensor

The stress tensor represents an *external exertion* on the material, while the strain tensor represents the *material reaction* to that exertion. In crystallography these are called **field tensors** because they represent imposed conditions, opposed to **matter tensors**, that represents material properties.

Hooke's law says that, for small deformations, stress is related to strain by a matter tensor called **elasticity tensor** or **stiffness tensor** E:

$$\sigma^{ij} = E^{ijk\ell}\varepsilon_{k\ell},$$

while the tensor relating strain to stress is the **compliance tensor** S:

$$\varepsilon_{k\ell} = S_{ijk\ell}\sigma^{ij}.$$

The elasticity tensor has order 4, and hence in 3-dimensional space it has $3^4 = 81$ components. Luckily, symmetry reduces the number of *independent* components for $E^{ijk\ell}$.

(1) **Minor Symmetries**

The symmetry of the stress tensor

$$\sigma^{ij} = \sigma^{ji}$$

implies that

$$E^{ijk\ell} = E^{jik\ell} \quad \text{for each } k, \ell;$$

it follows that for each k, ℓ fixed there are only 6 independent components $E^{ijk\ell}$

$$\begin{pmatrix} E^{11k\ell} & E^{12k\ell} & E^{13k\ell} \\ E^{12k\ell} & E^{22k\ell} & E^{23k\ell} \\ E^{13k\ell} & E^{23k\ell} & E^{33k\ell} \end{pmatrix}.$$

Having taken this in consideration, the number of independent components decreases to 6×3^2 at the most. Moreover, the symmetry also of the strain tensor

$$\varepsilon_{k\ell} = \varepsilon_{\ell k}$$

implies that

$$E^{ijk\ell} = E^{ij\ell k} \quad \text{for each } i, j.$$

This means that for each i, j fixed there are also only 6 independent components $E^{ijk\ell}$, so that $E^{ijk\ell}$ has at most $6^2 = 36$ independent components.

(2) **Major Symmetries**

Since (under appropriate conditions) partial derivatives commute, if follows from the existence of a *strain energy density functional U* satisfying

$$\frac{\partial^2 U}{\partial \varepsilon_{ij} \partial \varepsilon_{k\ell}} = E^{ijk\ell}$$

that

$$E^{ijk\ell} = E^{k\ell ij},$$

that means the matrix with rows labelled by (i, j) and columns labelled by (k, ℓ) is symmetric. Since, according to the previous minor symmetries, there are only 6 entries (i, j) for a fixed (k, ℓ), $E^{ijk\ell}$ can be written in a 6×6 matrix with rows labelled by (i, j) and columns labelled by (k, ℓ)

$$\begin{pmatrix} * & * & * & * & * & * \\ & * & * & * & * & * \\ & & * & * & * & * \\ & & & * & * & * \\ & & & & * & * \\ & & & & & * \end{pmatrix}$$

so that $E^{ijk\ell}$ has in fact only $6 + 5 + 4 + 3 + 2 + 1 = 21$ components.

6.5 Conductivity Tensor

Consider a homogeneous continuous crystal. Its properties can be divided into two classes:

- Properties that *do not depend* on a direction, and are hence described by *scalars*. Examples are density and heat capacity.
- Properties that *depend* on a direction, and are hence described by *tensors*. Examples are **elasticity**, **electrical conductivity** and **heat conductivity**. We say that a crystal is **anisotropic** when it has such actual *tensorial* properties.

6.5.1 Electrical Conductivity

Let E be the **electric field** and J the **electrical current density**. At each point of the crystal:

(1) E gives the **electric force** (in V/m, i.e., volt per meter) that would be exerted on a positive test charge (of 1 coulomb) placed at the point;
(2) J (in A/m^2 where A denotes the ampere unit) gives the direction the charge carriers move and the **rate of electric current** across an infinitesimal surface perpendicular to that direction.

The electrical current density J is a function of the electric field E,

$$J = f(E).$$

Consider a small increment ΔJ in J caused by a small increment ΔE in E, and write these increments in terms of their components with respect to a chosen orthonormal basis $\{e_1, e_2, e_3\}$.

$$\Delta J = \Delta J^i e_i \quad \text{and} \quad \Delta E = \Delta E^i e_i.$$

By Calculus, the increments are related by

$$\Delta J^i = \frac{\partial f^i}{\partial E^j} \Delta E^j + \text{higher order terms in } (\Delta E^j)^2, (\Delta E^j)^3, \ldots$$

If the quantities ΔE^j are small, we can assume that

$$\Delta J^i = \frac{\partial f^i}{\partial E^j} \Delta E^j \qquad (6.13)$$

If we assume that $\frac{\partial f^i}{\partial E^j}$ is independent of the point of the crystal,

$$\frac{\partial f^i}{\partial E^j} = \kappa^i_j \in \mathbb{R}$$

(in $\Omega^{-1}/\text{m}^{-1}$ where $\Omega = \text{V A}^{-1}$ is the ohm unit of resistance) we obtain the relation

$$\Delta J^i = \kappa^i_j \Delta E^j$$

or simply

$$\Delta J = \kappa \, \Delta E \, ,$$

where κ if the **electrical conductivity tensor**, sometimes denoted σ or γ. This is a $(1, 1)$-tensor and may depend[6] on the initial value of E, that is the electrical conductivity may be different for small and large electric forces. If initially $E = 0$ and κ^0 is the corresponding electrical conductivity tensor, we obtain the relation

$$J = \kappa^0 E$$

that is called the *generalized Ohm law*. This is always under the assumption that ΔE and ΔJ are small and that the relation is linear.

The **electrical resistivity tensor** (in $\Omega\,\text{m}$) is the inverse of κ:

$$\rho := \kappa^{-1} ,$$

that is, it is the $(1, 1)$-tensor such that

$$\rho^j_i \kappa^\ell_j = \delta^\ell_i .$$

The electrical conductivity measures the material's ability to conduct an electrical current, while the electrical resistivity quantifies the ability of the material to oppose the flow of the electrical current.

For an *isotropic* crystal, all directions are equivalent and these tensors are *spherical*, meaning

$$\kappa^j_i = k \delta^j_i \quad \text{and} \quad \rho^j_i = \frac{1}{k} \delta^j_i , \tag{6.14}$$

---

[6] Typically, if the dependence between E and J is linear for any value, and not only for small ones, the tensor κ will not depend on the initial value of E.

6.5 Conductivity Tensor

where k is a scalar, called the **electrical conductivity** of the crystal. Equation 6.14 can also be written as

$$\begin{pmatrix} k & 0 & 0 \\ 0 & k & 0 \\ 0 & 0 & k \end{pmatrix} \quad \text{and} \quad \begin{pmatrix} \frac{1}{k} & 0 & 0 \\ 0 & \frac{1}{k} & 0 \\ 0 & 0 & \frac{1}{k} \end{pmatrix}.$$

In general, κ_i^j is neither symmetric nor antisymmetric (and actually *symmetry* does not even make sense for a $(1, 1)$ tensor unless a metric is fixed, since it does require a canonical identification of V with V^*).

6.5.2 Heat Conductivity

Let T be the *temperature* and H the *heat flux vector*. For a homogeneous crystal, with constant H and for a constant gradient of T, the *Fourier Heat Conduction Law* asserts that

$$H = -K \operatorname{grad} T. \tag{6.15}$$

At each point of the crystal:

(1) $\operatorname{grad} T$ points in the direction of the highest ascent of the temperature and measures the rate of increase of T in that direction. The minus sign in Eq. 6.15 comes from the fact that the heat flows in the direction of decreasing temperature. Recall that the gradient of a real function is a covariant 1-tensor (Exercise 4.33).
(2) H measures the amount of heat passing per unit area perpendicular to its direction per unit time.

Here, K is the **heat conductivity tensor** or **thermal conductivity tensor**. In terms of components with respect to a chosen orthonormal basis, we have

$$H^i = -K^{ij}(\operatorname{grad} T)_j.$$

The heat conductivity tensor is a contravariant 2-tensor and experiments show that it is symmetric and hence can be orthogonally diagonalized. With respect to an orthonormal eigenbasis, K is represented by

$$\begin{pmatrix} K_1 & 0 & 0 \\ 0 & K_2 & 0 \\ 0 & 0 & K_3 \end{pmatrix},$$

where the eigenvalues of K are called the **principal coefficients** of heat conductivity. The eigenspaces of K are called the **principal directions**. The fact that heat flows always in the direction of decreasing temperature shows that the eigenvalues are positive

$$K_i > 0,$$

so, in particular, K is invertible. The **heat resistivity tensor** is the inverse of the heat conductivity tensor:

$$r := K^{-1},$$

and hence is also symmetric.

Solutions to Exercises

Chapter 2

EXERCISE 2.5. (1) yes; (2) no; (3) no; (4) yes; (5) no.

EXERCISE 2.8. (1) yes; (2) no; (3) no.

EXERCISE 2.9. Addition of linear transformations $T_1 : \mathbb{R}^2 \to \mathbb{R}^3$ and $T_2 : \mathbb{R}^2 \to \mathbb{R}^3$, and their multiplication by a scalar $\alpha \in \mathbb{R}$ are defined pointwise:

$$(T_1+T_2)(v) := T_1(v)+T_2(v) \quad \text{and} \quad (\alpha T_1)(v) := \alpha\,(T_1(v)) \quad \text{for each } v \in \mathbb{R}^2.$$

We check the required properties in Definition 2.1:

(1) and (2) follow pointwise from associativity and commutativity of sum of vectors in $\mathbb{R}^3$.
(3) is verified by the zero transformation, which sends all vectors in $\mathbb{R}^2$ to the zero vector of $\mathbb{R}^3$.
The additive inverse of a transformation $T : \mathbb{R}^2 \to \mathbb{R}^3$ is the transformation $-T$, which sends a vector v to the negative of $T(v)$, thus yielding (4).
(5)–(8) also follow pointwise from the corresponding properties of vectors in $\mathbb{R}^3$.

EXERCISE 2.12. (1) yes; (2) yes; (3) no.

EXERCISE 2.14. We first check the conditions for the kernel:

(1)' ker T is non-empty because it always contains at least the zero vector of V, since any linear transformation satisfies $T(0) = 0$;
(2)' ker T is closed under linear combinations, since, if $v_1, v_2 \in \ker T$, i.e., $T(v_1) = T(v_2) = 0$, and $\alpha, \beta \in \mathbb{R}$, then by linearity of T we have that

$$T(\alpha v_1 + \beta v_2) = \alpha T(v_1) + \beta T(v_2) = 0 + 0 = 0,$$

showing that $\alpha v_1 + \beta v_2 \in \ker T$.

Now we check the conditions for the image:

(1)' im T is non-empty because it always contains at least the zero vector of W, which is the image of the zero vector of V;
(2)' im T is closed under linear combinations, since, if $w_1, w_2 \in \operatorname{im} T$, i.e., there exist $v_1, v_2 \in V$ such that $T(v_1) = w_1$ and $T(v_2) = w_2$, and $\alpha, \beta \in \mathbb{R}$, then by linearity of T we have that

$$T(\alpha v_1 + \beta v_2) = \alpha T(v_1) + \beta T(v_2) = \alpha w_1 + \beta w_2,$$

showing that $\alpha w_1 + \beta w_2 \in \operatorname{im} T$.

EXERCISE 2.21. Let $W = \operatorname{span}\{b_1, \ldots, b_n\} = \{\alpha_1 b_1 + \alpha_2 b_2 + \cdots \alpha_n b_n : \alpha_1, \ldots \alpha_n \in \mathbb{R}\}$ denote the set of all linear combinations of the vectors $b_1, b_2 \ldots b_n \in V$. To show that W is a subspace of V we need to show that $0 \in W$ and W is closed under linear combinations.

(1) $0 \in W$: Since $0 = 0 b_1 + 0 b_2 + \cdots 0 b_n$, we indeed have that $0 \in W$ and W is non-empty.
(2) W is closed under linear combinations:
  Let $u, w \in W$ and $\gamma, \delta \in \mathbb{R}$. Then $u = \alpha_1 b_1 + \alpha_2 b_2 + \cdots \alpha_n b_n$ for some $\alpha_1, \ldots \alpha_n \in \mathbb{R}$ and $w = \beta_1 b_1 + \beta_2 b_2 + \cdots \beta_n b_n$ for some $\beta_1, \ldots \beta_n \in \mathbb{R}$. Then $\gamma u + \delta w = (\gamma \alpha_1 + \delta \beta_1) b_1 + \cdots (\gamma \alpha_n + \delta \beta_n) b_n$, being a linear combination of the vectors $b_1, b_2, \ldots b_n$, is again in W.

EXERCISE 2.29.

(1) Since these are 3 vectors in $\mathbb{R}^3$ and $\dim \mathbb{R}^3 = 3$, it is enough to show that these vectors are linearly independent. This amounts to showing that the only solution of the equation

$$\mu_1\begin{bmatrix}1\\0\\0\end{bmatrix}+\mu_2\begin{bmatrix}1\\1\\0\end{bmatrix}+\mu_3\begin{bmatrix}1\\1\\1\end{bmatrix}=\begin{bmatrix}0\\0\\0\end{bmatrix} \quad \text{i.e.} \quad \begin{bmatrix}1&1&1\\0&1&1\\0&0&1\end{bmatrix}\begin{bmatrix}\mu_1\\\mu_2\\\mu_3\end{bmatrix}=\begin{bmatrix}0\\0\\0\end{bmatrix}$$

is the trivial solution (given by $\mu_1 = \mu_2 = \mu_3 = 0$), which can be verified with Gauss-Jordan elimination, for instance.

(2) The coordinate vector,

$$[v]_\mathcal{B} = \begin{bmatrix}\mu_1\\\mu_2\\\mu_3\end{bmatrix},$$

is the solution of the equation

$$\mu_1\begin{bmatrix}1\\0\\0\end{bmatrix}+\mu_2\begin{bmatrix}1\\1\\0\end{bmatrix}+\mu_3\begin{bmatrix}1\\1\\1\end{bmatrix}=v \quad \text{i.e.} \quad \begin{bmatrix}1&1&1\\0&1&1\\0&0&1\end{bmatrix}\begin{bmatrix}\mu_1\\\mu_2\\\mu_3\end{bmatrix}=\begin{bmatrix}0\\1\\\pi\end{bmatrix}$$

Using Gauss-Jordan elimination, we find

$$[v]_\mathcal{B} = \begin{bmatrix}-1\\1-\pi\\\pi\end{bmatrix},$$

(3) By definition of coordinate vector, the vector w is

$$w = \begin{bmatrix}1\\0\\0\end{bmatrix}+2\begin{bmatrix}1\\1\\0\end{bmatrix}+3\begin{bmatrix}1\\1\\1\end{bmatrix}=\begin{bmatrix}6\\5\\3\end{bmatrix}.$$

EXERCISE 2.30.

(1) The vectors in $\mathcal{B}$ span V, since any element of V is of the form

$$\begin{bmatrix}a&b\\c&-a\end{bmatrix}=a\begin{bmatrix}1&0\\0&-1\end{bmatrix}+b\begin{bmatrix}0&1\\0&0\end{bmatrix}+c\begin{bmatrix}0&0\\1&0\end{bmatrix}.$$

Moreover, the vectors in $\mathcal{B}$ are linearly independent since

$$a\begin{bmatrix}1&0\\0&-1\end{bmatrix}+b\begin{bmatrix}0&1\\0&0\end{bmatrix}+c\begin{bmatrix}0&0\\1&0\end{bmatrix}=\begin{bmatrix}0&0\\0&0\end{bmatrix}$$

if and only if
$$\begin{bmatrix} a & b \\ c & -a \end{bmatrix} = \begin{bmatrix} 0 & 0 \\ 0 & 0 \end{bmatrix},$$
that is, if and only if $a = b = c = 0$.

(2) We know that $\dim V = 3$, as $\mathcal{B}$ is a basis of V and has three elements. Since $\widetilde{\mathcal{B}}$ also has three elements, it is enough to check either that it spans V or that it consists of linearly independent vectors. We will check this last condition. Indeed,
$$a \begin{bmatrix} 1 & 0 \\ 0 & -1 \end{bmatrix} + b \begin{bmatrix} 0 & -1 \\ 1 & 0 \end{bmatrix} + c \begin{bmatrix} 0 & 1 \\ 1 & 0 \end{bmatrix} = \begin{bmatrix} 0 & 0 \\ 0 & 0 \end{bmatrix} \iff \begin{bmatrix} a & c-b \\ b+c & -a \end{bmatrix} = \begin{bmatrix} 0 & 0 \\ 0 & 0 \end{bmatrix},$$
that is,
$$\begin{cases} a = 0 \\ b + c = 0 \\ c - b = 0 \end{cases} \iff \begin{cases} a = 0 \\ b = 0 \\ c = 0. \end{cases}$$

(3) Since
$$\begin{bmatrix} 2 & 1 \\ 7 & -2 \end{bmatrix} = 2 \begin{bmatrix} 1 & 0 \\ 0 & -1 \end{bmatrix} + 1 \begin{bmatrix} 0 & 1 \\ 0 & 0 \end{bmatrix} + 7 \begin{bmatrix} 0 & 0 \\ 1 & 0 \end{bmatrix},$$
we have
$$[v]_\mathcal{B} = \begin{pmatrix} 2 \\ 1 \\ 7 \end{pmatrix}.$$

To compute the coordinates of v with respect to $\widetilde{\mathcal{B}}$ we need to find $a, b, c \in \mathbb{R}$ such that
$$\begin{bmatrix} 2 & 1 \\ 7 & -2 \end{bmatrix} = a \begin{bmatrix} 1 & 0 \\ 0 & -1 \end{bmatrix} + b \begin{bmatrix} 0 & -1 \\ 1 & 0 \end{bmatrix} + c \begin{bmatrix} 0 & 1 \\ 1 & 0 \end{bmatrix}.$$

Solving the corresponding system of linear equations as above yields

$$[v]_{\widetilde{\mathcal{B}}} = \begin{pmatrix} 2 \\ 3 \\ 4 \end{pmatrix}.$$

EXERCISE 2.32.

(1) Let $C := AB \in \mathbb{R}^{\ell \times n}$. By definition, we have

$$C_k^i = A_1^i B_k^1 + A_2^i B_k^2 + \cdots + A_m^i B_k^m = \sum_{J=1}^{m} A_J^i B_k^J = A_j^i B_k^j.$$

(2) The column vector $By \in \mathbb{R}^{m \times 1}$ has coordinates

$$(By)^i = B_1^i y^1 + \cdots + B_n^i y^n = \sum_{J=1}^{n} B_J^i y^J = B_j^i y^j.$$

(3) The row vector $y^T \in \mathbb{R}^{1 \times n}$ has coordinates $(y^T)_i = y^i$. The transpose of B has entries $(B^T)_j^k = B_k^j$, where $1 \leq j \leq m$ and $1 \leq k \leq n$. Then we have

$$(y^T B^T)_j = \sum_{K=1}^{n} (y^T)_K (B^T)_j^K = (y^T)_k (B^T)_j^k = y^k B_k^j.$$

Alternatively, we may note that $y^T B^T = (By)^T$ and conclude that

$$(y^T B^T)_i = ((By)^T)_i = (By)^i = B_j^i y^j.$$

These two expressions for $y^T B^T$ are equivalent.

(4) We have $xy^T B^T \in \mathbb{R}^{\ell \times m}$, where $x \in \mathbb{R}^{\ell \times 1}$ has coordinates x^i and $y^T B^T \in \mathbb{R}^{1 \times m}$ has coordinates $(y^T B^T)_j = B_k^j y^k$. Therefore, we have

$$(xy^T B^T)_j^i = x^i B_k^j y^k.$$

EXERCISE 2.33. Let $\mathcal{A} = \{a_1, a_2, \ldots, a_n\}$, $\mathcal{B} = \{b_1, b_2, \ldots, b_n\}$ and $\mathcal{C} = \{c_1, c_2, \cdots, c_n\}$ be three bases of a vector space V. Let $L_{\mathcal{AB}} = (L_j^i)$ be the matrix of the change of basis from $\mathcal{A}$ to $\mathcal{B}$, then $b_j = L_j^i a_i$. Similary if $L_{\mathcal{BC}} =$

(M_j^i) is the matrix of the change of basis from $\mathcal{B}$ to $\mathcal{C}$, then $c_k = M_k^j b_j$. Putting these together gives

$$c_k = M_k^j b_j = M_k^j L_j^i a_i.$$

On the other hand, as $L_{\mathcal{AC}} = (N_j^i)$ is the change of basis matrix from $\mathcal{A}$ to $\mathcal{C}$, we also have that $c_k = N_k^i a_i$. Since any vector is a unique linear combination of the basis vectors $\{a_1, a_2 \ldots, a_n\}$, we must have that

$$M_k^j L_j^i = N_k^i.$$

But this is exactly the matrix product $L_{\mathcal{AB}} L_{\mathcal{BC}} = L_{\mathcal{AC}}$ written in the Einstein notation.

EXERCISE 2.38. We have that $(A^T)_j^i = A_i^j$, where $1 \leq i \leq \ell$ and $1 \leq j \leq m$. Then $(A^T x)^i = (A^T)_j^i x^j = \sum_{J=1}^m A_i^J x^J$. In the last expression, both J-indices are upper. In order to follow the Einstein convention, we use the Kronecker symbol, δ_{jk}. By definition, we have $\delta_{jk} = 1$ for $j = k$ und $\delta_{jk} = 0$ otherwise. Therefore, we have

$$(A^T x)^i = (A^T)_j^i x^j = \sum_{J=1}^m A_i^J x^J = \sum_{J=1}^m \sum_{K=1}^m A_i^J \delta_{JK} x^K = A_i^j \delta_{jk} x^k.$$

EXERCISE 2.50.

(1) The function α is indeed well-defined, since the derivative of a polynomial p of degree at most 3 is a polynomial of degree at most 2, hence the product $(x - 1)p'(x)$ is a polynomial of degree at most 3. Since differentiation is a linear transformation, we have that, for all $p, q \in V$ and $\lambda \in \mathbb{R}$,

$$\alpha(p + \lambda q) = (x - 1)(p + \lambda q)' = (x - 1)(p' + \lambda q')$$
$$= (x - 1)p' + \lambda(x - 1)q' = \alpha(p) + \lambda \alpha(q).$$

Therefore, α is a linear transformation.

(2) For each $k \in \{1, 2, 3\}$, we have that

$$\alpha((x - 1)^k) = (x - 1) \cdot k(x - 1)^{k-1} = k(x - 1)^k,$$

hence, $(x - 1)^k$ is an eigenvector with eigenvalue k. Moreover, we have that $\alpha(1) = 0$, showing that the constant polynomial 1 is an eigenvector with eigenvalue 0. Since they are eigenvectors with different eigenvalues, the elements of $\widetilde{\mathcal{B}}$ are linearly independent. Since there are four of them, they must form a basis of the four-dimensional vector space V, hence $\widetilde{\mathcal{B}}$ is an eigenbasis with respect to α.

(3) This is the diagonal matrix with the eigenvalues of α along the diagonal, in the order corresponding to that of the eigenvectors in the basis $\widetilde{\mathcal{B}}$:

$$\widetilde{M} = \begin{bmatrix} 0 & 0 & 0 & 0 \\ 0 & 1 & 0 & 0 \\ 0 & 0 & 2 & 0 \\ 0 & 0 & 0 & 3 \end{bmatrix}.$$

(4) It is understood that we work with the standard basis $\mathcal{E} = \{1\}$ of $\mathbb{R}$. The matrix of the linear transformation β with respect to the basis $\mathcal{B}$, resp. $\widetilde{\mathcal{B}}$, of V (and the basis $\mathcal{E} = \{1\}$ of $\mathbb{R}$) is obtained by computing the images of the basis elements:

$$\beta(1) = 1, \qquad \beta(x^n) = 1, \qquad \beta((x-1)^n) = 0.$$

From this we obtain the requested matrices, resp.:

$$A = \begin{bmatrix} 1 & 1 & 1 & 1 \end{bmatrix} \quad \text{and} \quad \widetilde{A} = \begin{bmatrix} 1 & 0 & 0 & 0 \end{bmatrix}.$$

(5) We write the (new) basis elements from $\widetilde{\mathcal{B}}$ in terms of the (old) basis elements from $\mathcal{B}$:

$$1 = 1,$$
$$x - 1 = -1 + x,$$
$$(x-1)^2 = 1 - 2x + x^2,$$
$$(x-1)^3 = -1 + 3x - 3x^2 + x^3.$$

By reading off the coefficients on the right-hand sides, and writing them columnwise, we obtain the first change of basis matrix:

$$L_{\mathcal{B}\widetilde{\mathcal{B}}} = \begin{bmatrix} 1 & -1 & 1 & -1 \\ 0 & 1 & -2 & 3 \\ 0 & 0 & 1 & -3 \\ 0 & 0 & 0 & 1 \end{bmatrix}.$$

The reverse change of basis matrix may be obtained by inverting the above $L_{B\widetilde{B}}$ or again by the above procedure:

$$1 = 1$$
$$x = 1 + (x - 1)$$
$$x^2 = (1 + x - 1)^2 = 1 + 2(x - 1) + (x - 1)^2$$
$$x^3 = (1 + x - 1)^3 = 1 + 3(x - 1) + 3(x - 1)^2 + (x - 1)^3,$$

yields that

$$L_{\widetilde{B}B} = L_{B\widetilde{B}}^{-1} = \begin{bmatrix} 1 & 1 & 1 & 1 \\ 0 & 1 & 2 & 3 \\ 0 & 0 & 1 & 3 \\ 0 & 0 & 0 & 1 \end{bmatrix}.$$

(6) Indeed, we have that

$$AL_{B\widetilde{B}} = \begin{bmatrix} 1 & 1 & 1 & 1 \end{bmatrix} \begin{bmatrix} 1 & -1 & 1 & -1 \\ 0 & 1 & -2 & 3 \\ 0 & 0 & 1 & -3 \\ 0 & 0 & 0 & 1 \end{bmatrix} = \begin{bmatrix} 1 & 0 & 0 & 0 \end{bmatrix} = \widetilde{A}.$$

Chapter 3

EXERCISE 3.2. (1) no; (2) no; (3) yes.

EXERCISE 3.3. (1) yes; (2) yes; (3) no.

EXERCISE 3.8. We show that V^* is a subspace of the vector space of all real-valued functions on V (cf. Example 2.3(3) in Sect. 2.1.1), by checking the three conditions:

(1) The 0-function **0** associating the number zero to each vector in V is linear because $0 + 0 = 0$ and $k0 = 0$ for every $k \in \mathbb{R}$, so $\mathbf{0} \in V^*$;
(2) V^* is *closed under addition* since, if $\alpha : V \to \mathbb{R}$ and $\beta : V \to \mathbb{R}$ are linear, then $\alpha + \beta : V \to \mathbb{R}$ defined by $(\alpha + \beta)(v) = \alpha(v) + \beta(v)$ is also linear (in $v \in V$);

(3) V^* is *closed under multiplication by scalars* since, if $\alpha : V \to \mathbb{R}$ is linear and $k \in \mathbb{R}$, then $k\alpha : V \to \mathbb{R}$ defined by $(k\alpha)(v) = k(\alpha(v))$ is also linear.

EXERCISE 3.13. In fact:
$$[\alpha]_{\mathcal{B}^*} L = (\alpha_1 \ldots \alpha_n) \begin{bmatrix} L_1^1 & \ldots & L_n^1 \\ \vdots & & \vdots \\ L_1^n & \ldots & L_n^n \end{bmatrix}$$
$$= (\alpha_i L_1^i \ldots \alpha_i L_n^i)$$
$$= (\widetilde{\alpha}_1 \ldots \widetilde{\alpha}_n)$$
$$= [\alpha]_{\widetilde{\mathcal{B}}^*}$$

EXERCISE 3.18. By the Laplace expansion formula, we have
$$\det \begin{bmatrix} u \\ v \\ w \end{bmatrix} = \det \begin{bmatrix} u^1 & u^2 & u^3 \\ v^1 & v^2 & v^3 \\ w^1 & w^2 & w^3 \end{bmatrix}$$
$$= u^1 \left(v^2 w^3 - v^3 w^2\right) + u^2 \left(v^3 w^1 - v^1 w^3\right) + u^3 \left(v^1 w^2 - v^2 w^1\right)$$
$$= \begin{bmatrix} u^1 \\ u^2 \\ u^3 \end{bmatrix} \cdot \begin{bmatrix} v^2 w^3 - v^3 w^2 \\ v^3 w^1 - v^1 w^3 \\ v^1 w^2 - v^2 w^1 \end{bmatrix} = u \cdot (v \times w).$$

EXERCISE 3.20. (1) yes; (2) yes; (3) yes; (4) no, because $v \times w$ is not a real number; (5) no, because it fails linearity (the area of the parallelogram spanned by v and w is the *same* as that of the parallelogram spanned by $-v$ and w); (6) no, because because it fails linearity in the second argument (the determinant of a matrix with $n > 1$ is not linear in that matrix).

EXERCISE 3.22.

(1) Indeed, $\varphi(x, y) := 2x - y$ is not linear in x (nor linear in y):
$$\varphi(ax + bz, y) = 2(ax + bz) - y$$
$$\neq a(2x - y) + b(2z - y)$$
$$= a\varphi(x, y) + b\varphi(z, y).$$

Yet, φ is linear in $\mathbb{R}^2$:

$$\varphi\left(a(x,y) + a'(x',y')\right) = \varphi\left(ax + a'x', ay + a'y'\right)$$
$$= 2(ax + a'x') - (ay + a'y')$$
$$= a(2x - y) + a'(2x' - y')$$
$$= a\varphi(x,y) + a'\varphi(x',y').$$

(2) Indeed, $\varphi(x,y) := 2xy$ is linear in x:

$$\varphi(ax + bz, y) = 2(ax + bz)y$$
$$= a(2xy) + b(2zy) = a\varphi(x,y) + b\varphi(z,y),$$

and, similarly, it is linear in y, hence a bilinear form on $\mathbb{R}$. On the other hand, φ is not linear in $(x,y) \in \mathbb{R}^2$:

$$\varphi\left(a(x,y) + a'(x',y')\right) = \varphi\left(ax + a'x', ay + a'y'\right)$$
$$= 2(ax + a'x')(ay + a'y')$$
$$\neq a(2xy) + a'(2x'y')$$
$$= a\varphi(x,y) + a'\varphi(x',y').$$

EXERCISE 3.26. Since $\mathrm{Bil}(V \times V, \mathbb{R})$ is a subset of the vector space $\{f : V \times V \to \mathbb{R}\}$ of all real-valued functions on the vector space $V \times V$, it is enough to check that it is a subspace of $\{f : V \times V \to \mathbb{R}\}$ Therefore, it is enough to check the two conditions:

(1) The zero function associating the number zero to each element from $V \times V$ is trivially bilinear, hence it is in $\mathrm{Bil}(V \times V, \mathbb{R})$.
(2) The subset $\mathrm{Bil}(V \times V, \mathbb{R})$ is closed under linear combinations. Assuming that $\varphi, \psi \in \mathrm{Bil}(V \times V, \mathbb{R})$, and $\lambda, \mu \in \mathbb{R}$, we check that $\lambda\varphi + \mu\psi$ is bilinear. The linearity in the first entry amounts to:

$$(\lambda\varphi + \mu\psi)(ax + bz, y) = \lambda\varphi(ax + bz, y) + \mu\psi(ax + bz, y)$$
$$= a\lambda\varphi(x,y) + b\lambda\varphi(z,y) + a\mu\psi(x,y) + b\mu\psi(z,y)$$
$$= a\left(\lambda\varphi(x,y) + \mu\psi(x,y)\right) + b\left(\lambda\varphi(z,y) + \mu\psi(z,y)\right)$$
$$= a(\lambda\varphi + \mu\psi)(x,y) + b(\lambda\varphi + \mu\psi)(z,y)$$

The linearity in the second entry is analogous.

7 Solutions to Exercises

EXERCISE 3.30. We have that

$$
{}^tLBL = \begin{bmatrix} L_1^1 & \dots & L_1^n \\ \vdots & & \vdots \\ L_n^1 & \dots & L_n^n \end{bmatrix} \begin{bmatrix} B_{11} & \dots & B_{1n} \\ \vdots & & \vdots \\ B_{n1} & \dots & B_{nn} \end{bmatrix} \begin{bmatrix} L_1^1 & \dots & L_n^1 \\ \vdots & & \vdots \\ L_1^n & \dots & L_n^n \end{bmatrix}
$$

$$
= \begin{bmatrix} L_1^1 & \dots & L_1^n \\ \vdots & & \vdots \\ L_n^1 & \dots & L_n^n \end{bmatrix} \begin{bmatrix} B_{1i} L_1^i & \dots & B_{1i} L_n^i \\ \vdots & & \vdots \\ B_{ni} L_1^i & \dots & B_{ni} L_n^i \end{bmatrix}
$$

$$
= \begin{bmatrix} L_1^k B_{ki} L_1^i & \dots & L_1^k B_{ki} L_n^i \\ \vdots & & \vdots \\ L_n^k B_{ki} L_1^i & \dots & L_n^k B_{ki} L_n^i \end{bmatrix}
$$

$$
= \widetilde{B}.
$$

Chapter 4

EXERCISE 4.2. We first check that $\varphi(v, w) := v \cdot w$ is a bilinear form on $\mathbb{R}^3$:

$$\varphi(av + bz, w) = (av + bz) \cdot w = (av^i + bz^i) w^j \delta_{ij}$$
$$= a(v^i w^j \delta_{ij}) + b(z^i w^j \delta_{ij}) = a(v \cdot w) + b(z \cdot w)$$
$$= a\varphi(v, w) + b\varphi(z, w)$$

and similarly for the linearity in the second entry. Now, we check that φ is symmetric

$$v \cdot w = v^i w^j \delta_{ij} = w^i v^j \delta_{ij} = w \cdot v.$$

Finally, we check positive definiteness:

$$v \cdot v = v^i v^j \delta_{ij} = (v^1)(v^1) + (v^2)(v^2) + (v^3)(v^3) \text{ is always non-negative}$$

because it is a sum of squares, and vanishes if and only if all these squares vanish, that is, when $v = 0$.

EXERCISE 4.3.

(1) No, as φ is *negative definite*, that is $\varphi(v,v) < 0$ if $v \in V$, $v \neq 0$.
(2) No, as φ is not symmetric.
(3) No, as φ is not positive definite.
(4) No, as φ is not positive definite.
(5) Yes.
(6) Yes.

EXERCISE 4.4.

(1) Yes, in fact:
   (a) $\int_0^1 p(x)q(x)dx = \int_0^1 q(x)p(x)dx$ because $p(x)q(x) = q(x)p(x)$;
   (b) $\int_0^1 (p(x))^2 dx \geq 0$ for all $p \in P_2(\mathbb{R})$ because $(p(x))^2 \geq 0$, and $\int_0^1 (p(x))^2 dx = 0$ only when $p(x) = 0$ for all $x \in [0,1]$, that is only if $p \equiv 0$.
(2) No, since $\int_0^1 (p'(x))^2 dx = 0$ implies that $p'(x) = 0$ for all $x \in [0,1]$, but such p is not necessarily the zero polynomial.
(3) Yes.
(4) No. Is there $p \in P_2(\mathbb{R})$, $p \neq 0$ such that $(p(1))^2 + (p(2))^2 = 0$?
(5) Yes. Is there a non-zero polynomial of degree 2 with 3 distinct zeros?
(6) No, since this is not symmetric.

EXERCISE 4.12. We write

$$[v]_{\widetilde{\mathcal{B}}} = \begin{pmatrix} \tilde{v}^1 \\ \tilde{v}^2 \\ \tilde{v}^3 \end{pmatrix} \quad \text{and} \quad [w]_{\widetilde{\mathcal{B}}} = \begin{pmatrix} \tilde{w}^1 \\ \tilde{w}^2 \\ \tilde{w}^3 \end{pmatrix}.$$

We know that g with respect to the basis $\widetilde{\mathcal{B}}$ has the standard form $g(v,w) = \tilde{v}^i \tilde{w}^i$ and we want to verify the formula (4.7) using the matrix of the change of coordinates $L^{-1} = \Lambda$. If

$$[v]_{\mathcal{B}} = \begin{pmatrix} v^1 \\ v^2 \\ v^3 \end{pmatrix} \quad \text{and} \quad [w]_{\mathcal{B}} = \begin{pmatrix} w^1 \\ w^2 \\ w^3 \end{pmatrix}$$

then we have that

$$\begin{pmatrix} \tilde{v}^1 \\ \tilde{v}^2 \\ \tilde{v}^3 \end{pmatrix} = \Lambda \begin{pmatrix} v^1 \\ v^2 \\ v^3 \end{pmatrix} = \begin{pmatrix} v^1 - v^2 \\ v^2 - v^3 \\ v^3 \end{pmatrix}$$

7 Solutions to Exercises

and

$$\begin{pmatrix} \tilde{w}^1 \\ \tilde{w}^2 \\ \tilde{w}^3 \end{pmatrix} = \Lambda \begin{pmatrix} w^1 \\ w^2 \\ w^3 \end{pmatrix} = \begin{pmatrix} w^1 - w^2 \\ w^2 - w^3 \\ w^3 \end{pmatrix}$$

It follows that

$$g(v, w) = \tilde{v}^i \tilde{w}^i = (v^1 - v^2)(w^1 - w^2) + (v^2 - v^3)(w^2 - w^3) + v^3 w^3$$
$$= v^1 w^1 - v^1 w^2 - v^2 w^1 + 2v^2 w^2 - v^2 w^3 - w^3 v^2 + 2v^3 w^3.$$

EXERCISE 4.15. With respect to $\widetilde{\mathcal{B}}$, we have

$$\|v\| = (1^2 + 1^2 + 1^2)^{1/2} = \sqrt{3}$$
$$\|w\| = ((-1)^2 + (-1)^2 + 3^2)^{1/2} = \sqrt{11}$$

and with respect to $\mathcal{E}$

$$\|v\| = (3 \cdot 3 - 3 \cdot 2 - 2 \cdot 3 + 2 \cdot 2 \cdot 2 - 2 \cdot 1 - 1 \cdot 2 + 2 \cdot 1 \cdot 1)^{1/2} = \sqrt{3}$$
$$\|w\| = (1 \cdot 1 - 1 \cdot 2 - 2 \cdot 1 + 2 \cdot 2 \cdot 2 - 2 \cdot 3 - 3 \cdot 2 + 2 \cdot 3 \cdot 3)^{1/2} = \sqrt{11}.$$

EXERCISE 4.16. Saying that the orthogonality is meant with respect to g, means that we have to show that $g(v - \text{proj}_{b_k} v, b_k) = 0$. In fact,

$$g(v - \text{proj}_{b_k} v, b_k) = g(v - \frac{g(v, b_k)}{g(b_k, b_k)} b_k, b_k) = g(v, b_k) - \frac{g(v, b_k)}{g(b_k, b_k)} g(b_k, b_k) = 0$$

EXERCISE 4.21.

(1) The coordinate vectors of basis vectors with respect to that same basis are simply standard vectors, in this case:

$$[\tilde{b}_1]_{\widetilde{\mathcal{B}}} = \begin{pmatrix} 1 \\ 0 \\ 0 \end{pmatrix}, [\tilde{b}_2]_{\widetilde{\mathcal{B}}} = \begin{pmatrix} 0 \\ 1 \\ 0 \end{pmatrix} \text{ and } [\tilde{b}_3]_{\widetilde{\mathcal{B}}} = \begin{pmatrix} 0 \\ 0 \\ 1 \end{pmatrix}.$$

(2) As in Example 4.11, we have $G_{\widetilde{\mathcal{B}}} = I$ and $G_{\mathcal{E}} = \begin{bmatrix} 1 & -1 & 0 \\ -1 & 2 & -1 \\ 0 & -1 & 2 \end{bmatrix}$.

(3) In parts (a) and (b), note that, for an orthonormal basis $\tilde{\mathcal{B}}$, we have $\tilde{\mathcal{B}}^g = \tilde{\mathcal{B}}$. In parts (c) and (d), we use the computations in Example 4.20 and the fact that $[v]_{\mathcal{E}} = L_{\mathcal{E}\tilde{\mathcal{B}}}[v]_{\tilde{\mathcal{B}}}$.

(a) $[\tilde{b}^1]_{\tilde{\mathcal{B}}} = [\tilde{b}_1]_{\tilde{\mathcal{B}}} = \begin{pmatrix} 1 \\ 0 \\ 0 \end{pmatrix}$, $[\tilde{b}^2]_{\tilde{\mathcal{B}}} = [\tilde{b}_2]_{\tilde{\mathcal{B}}} = \begin{pmatrix} 0 \\ 1 \\ 0 \end{pmatrix}$ and $[\tilde{b}^3]_{\tilde{\mathcal{B}}} = [\tilde{b}_3]_{\tilde{\mathcal{B}}} = \begin{pmatrix} 0 \\ 0 \\ 1 \end{pmatrix}$.

(b) $[\tilde{b}^1]_{\mathcal{E}} = [\tilde{b}_1]_{\mathcal{E}} = \begin{pmatrix} 1 \\ 0 \\ 0 \end{pmatrix}$, $[\tilde{b}^2]_{\mathcal{E}} = [\tilde{b}_2]_{\mathcal{E}} = \begin{pmatrix} 1 \\ 1 \\ 0 \end{pmatrix}$ and $[\tilde{b}^3]_{\mathcal{E}} = [\tilde{b}_3]_{\mathcal{E}} = \begin{pmatrix} 1 \\ 1 \\ 1 \end{pmatrix}$.

(c) $[e^1]_{\tilde{\mathcal{B}}} = \begin{pmatrix} 1 \\ 1 \\ 1 \end{pmatrix}$, $[e^2]_{\tilde{\mathcal{B}}} = \begin{pmatrix} 0 \\ 1 \\ 1 \end{pmatrix}$ and $[e^3]_{\tilde{\mathcal{B}}} = \begin{pmatrix} 0 \\ 0 \\ 1 \end{pmatrix}$.

(d) $[e^1]_{\mathcal{E}} = \begin{pmatrix} 3 \\ 2 \\ 1 \end{pmatrix}$, $[e^2]_{\mathcal{E}} = \begin{pmatrix} 2 \\ 2 \\ 1 \end{pmatrix}$ and $[e^3]_{\mathcal{E}} = \begin{pmatrix} 1 \\ 1 \\ 1 \end{pmatrix}$.

EXERCISE 4.24.

(1) The assertion in the case of the bases $\mathcal{E}$ and $\mathcal{E}^g$ follows from

$$G_{\mathcal{E}}^{-1} = (L_{\mathcal{E}^g\mathcal{E}})^{-1} = \begin{bmatrix} | & | & | \\ e^1 & e^2 & e^3 \\ | & | & | \end{bmatrix} = \begin{bmatrix} 3 & 2 & 1 \\ 2 & 2 & 1 \\ 1 & 1 & 1 \end{bmatrix}.$$

(2) Since $\tilde{b}^j = \tilde{b}_j$, we have $G_{\tilde{\mathcal{B}}}^{-1} = L_{\tilde{\mathcal{B}}^g\tilde{\mathcal{B}}} = I$ and Eq. (4.15) is immediately verified.

EXERCISE 4.32. We have that

$$T^i(t) = L^i_j \tilde{T}^j(t),$$

where $T^i(t)$, respectively $\tilde{T}^j(t)$, are the components of T at time t with respect to a basis $\mathcal{B}$, respectively $\tilde{\mathcal{B}}$, and L^i_j are the components of the change of basis from $\mathcal{B}$ to $\tilde{\mathcal{B}}$. By differentiating the above equation with respect to time t, we obtain a similar equation for the components of $\frac{dT}{dt}$, namely

$$\frac{dT^i}{dt} = L^i_j \frac{d\widetilde{T}^j}{dt},$$

hence $\frac{dT}{dt}$ has the same transformation character as T.

A similar reasoning applies to arbitrary tensors.

EXERCISE 4.33. In order to discuss the partial derivatives, we express $f: V \to \mathbb{R}$ as a function of the corresponding coordinate-variables with respect to two bases, $\mathcal{B}$ and $\widetilde{\mathcal{B}}$, of V:

$$f(x) = f(x^i b_i) = f_\mathcal{B}(x^1, \ldots, x^n)$$

and

$$f(x) = f(\widetilde{x}^i \widetilde{b}_i) = f_{\widetilde{\mathcal{B}}}(\widetilde{x}^1, \ldots, \widetilde{x}^n)$$

Since $x^i = L^i_j \widetilde{x}^j$ where L is the change of basis from $\mathcal{B}$ to $\widetilde{\mathcal{B}}$, the partial derivatives satisfy

$$\begin{aligned}
\frac{\partial f}{\partial \widetilde{x}^j} &= \frac{\partial}{\partial \widetilde{x}^j} f_{\widetilde{\mathcal{B}}}(\widetilde{x}^1, \ldots, \widetilde{x}^n) \\
&= \frac{\partial}{\partial \widetilde{x}^j} f_\mathcal{B}(x^1, \ldots, x^n) \\
&= \frac{\partial}{\partial \widetilde{x}^j} f_\mathcal{B}(L^1_k \widetilde{x}^k, \ldots, L^n_k \widetilde{x}^k) \\
&= \left(\frac{\partial}{\partial x^i} f_\mathcal{B}(x^1, \ldots, x^n)\right) \cdot \underbrace{\left(\frac{\partial}{\partial \widetilde{x}^j} L^i_k \widetilde{x}^k\right)}_{=L^i_j} \\
&= L^i_j \frac{\partial f}{\partial x^i}.
\end{aligned}$$

We used the chain rule in the step before the last. The above transformation behaviour is that of a tensor of type (0, 1). We conclude that the gradient is a covector, i.e., a covariant 1-tensor.

EXERCISE 4.34. Let $\mathcal{B}$ and $\widetilde{\mathcal{B}}$ be two bases, with change of basis matrix $L = \Lambda^{-1}$. We write $\widetilde{T}$, $\widetilde{u}$ und $\widetilde{v}$ for the numbers or components with respect to the basis $\widetilde{\mathcal{B}}$. For the basis $\mathcal{B}$, we write these without tilde.

(1) We have that

$$\tilde{v}_j := \tilde{T}_{ij}\tilde{u}^i = L_i^{\ell_1} L_j^{\ell_2} T_{\ell_1 \ell_2} \Lambda_{k_1}^i u^{k_1} \quad \text{as } T \text{ is a } (0,2) \text{ and } u \text{ a } (1,0)\text{-tensor}$$

$$= L_i^{\ell_1} \Lambda_{k_1}^i L_j^{\ell_2} T_{\ell_1 \ell_2} u^{k_1} \quad \text{since multiplication is commutative}$$

$$= \delta_{k_1}^{\ell_1} L_j^{\ell_2} T_{\ell_1 \ell_2} u^{k_1} \quad \text{since } L \text{ is the inverse of } \Lambda$$

$$= L_j^{\ell_2} T_{\ell_1 \ell_2} u^{\ell_1} \quad \text{by definition of } \delta$$

$$= L_j^{\ell_2} v_{\ell_2}, \quad \text{showing, that } v \text{ is a } (0,1)\text{-tensor}.$$

(2) For arbitrary u^i, we have

$$\tilde{T}_{ij}\tilde{u}^i =: \tilde{v}_j = L_j^{\ell_2} v_{\ell_2} \quad \text{since } v \text{ is a } (0,1)\text{-tensor}$$

$$= L_j^{\ell_2} T_{\ell_1 \ell_2} u^{\ell_1} \quad \text{by definition of } v$$

$$= L_j^{\ell_2} T_{\ell_1 \ell_2} \delta_k^{\ell_1} u^k \quad \text{by definition of } \delta$$

$$= L_j^{\ell_2} T_{\ell_1 \ell_2} L_i^{\ell_1} \Lambda_k^i u^k \quad \text{since } L \text{ is the inverse of } \Lambda$$

$$= L_i^{\ell_1} L_j^{\ell_2} T_{\ell_1 \ell_2} \tilde{u}^i \quad \text{since } u \text{ is a } (1,0)\text{-tensor}.$$

As $\tilde{u}^i$ was arbitrary, we must indeed have

$$\tilde{T}_{ij} = L_i^{\ell_1} L_j^{\ell_2} T_{\ell_1 \ell_2}.$$

Chapter 5

EXERCISE 5.3. Since $\text{Bil}(V^* \times V^*, \mathbb{R})$ is a subset of the vector space $\{f : V^* \times V^* \to \mathbb{R}\}$ of all real-valued functions on the vector space $V^* \times V^*$, it is enough to check that it is a subspace of $\{f : V^* \times V^* \to \mathbb{R}\}$ Therefore, it is enough to check the two conditions:

(1) The zero function associating the number zero to each element from $V^* \times V^*$ is trivially bilinear, hence it is in $\text{Bil}(V^* \times V^*, \mathbb{R})$.

7 Solutions to Exercises

(2) The subset $\text{Bil}(V^* \times V^*, \mathbb{R})$ is closed under linear combinations. Assuming that $\sigma, \tau \in \text{Bil}(V^* \times V^*, \mathbb{R})$, and $c, d \in \mathbb{R}$, we check that $c\sigma + d\tau$ is bilinear. The linearity in the first entry amounts to:

$$(c\sigma + d\tau)(a\alpha + b\beta, \gamma) = c\sigma(a\alpha + b\beta, \gamma) + d\tau(a\alpha + b\beta, \gamma)$$
$$= ac\sigma(\alpha, \gamma) + bc\sigma(\beta, \gamma) + ad\tau(\alpha, \gamma)$$
$$+ bd\tau(\beta, \gamma)$$
$$= a\left(c\sigma(\alpha, \gamma) + d\tau(\alpha, \gamma)\right)$$
$$+ b\left(c\sigma(\beta, \gamma) + d\tau(\beta, \gamma)\right)$$
$$= a(c\sigma + d\tau)(\alpha, \gamma) + b(c\sigma + d\tau)(\beta, \gamma).$$

The linearity in the second entry is analogous.

EXERCISE 5.7. Let A and B be square matrices of the same size. If A has (i, j)-entry A^i_j (where i labels the row and j the column) and B has (i, j)-entry B^i_j, then by the definition of matrix product the matrix $C := AB$ has (i, j)-entry

$$A^i_k B^k_j$$

and the transpose of A has (i, j)-entry A^j_i, so $C^t A$ has (i, j)-entry

$$C^i_\ell A^j_\ell$$

and $AB^t A$ has (i, j)-entry

$$A^i_k B^k_\ell A^j_\ell.$$

We obtain Eq. (5.3) by replacing A by Λ and B^k_ℓ by $S^{k\ell}$.

EXERCISE 5.9. We set $T^{i,j,k} := T(\varepsilon^i, \varepsilon^j, \varepsilon^k)$. Then we have that

$$T^{i,j,1} = i + j \text{ and } T^{i,j,2} = i - j.$$

(1) The value of T on an arbitray triple

$$([\alpha_1 \; \alpha_2], [\beta_1 \; \beta_2], [\gamma_1 \; \gamma_2]) \in (V^*)^3$$

is defined by

$$T([\alpha_1\ \alpha_2],[\beta_1\ \beta_2],[\gamma_1\ \gamma_2]) = \alpha_i \beta_j \gamma_k T^{i,j,k},$$

where we have used the Einstein convention and $1 \leq i,j,k \leq 2$. Hence, we have

$$\begin{aligned}T([-1\ 2],[3\ 2],[1\ 1]) &= \alpha_1\beta_1\gamma_1 T^{1,1,1} + \alpha_2\beta_1\gamma_1 T^{2,1,1} + \alpha_1\beta_2\gamma_1 T^{1,2,1}\\&\quad + \alpha_1\beta_1\gamma_2 \underbrace{T^{1,1,2}}_{=0} + \alpha_1\beta_2\gamma_2 T^{1,2,2} + \alpha_2\beta_2\gamma_1 T^{2,2,1}\\&\quad + \alpha_2\beta_1\gamma_2 T^{2,1,2} + \alpha_2\beta_2\gamma_2 \underbrace{T^{2,2,2}}_{=0}\\&= -3\cdot 2 + 6\cdot 3 - 2\cdot 3 - 2\cdot(-1) + 4\cdot 4 + 6\cdot 1\\&= 30.\end{aligned}$$

(2) We apply the transformation formula from Sect. 5.2. The change of basis matrix from the standard basis to $\mathcal{B}$ is

$$L := L_{\mathcal{E}\mathcal{B}} = \begin{bmatrix} 1 & 1 \\ -1 & 0 \end{bmatrix}$$

with inverse

$$\Lambda := L^{-1} = \begin{bmatrix} 0 & -1 \\ 1 & 1 \end{bmatrix}.$$

We denote by $\widetilde{T}^{\ell,m,n}$ the (ℓ,m,n)-component of T with respect to $\mathcal{B}$, where $1 \leq \ell,m,n \leq 2$. Then we have

$$\widetilde{T}^{\ell,m,n} = \Lambda_i^\ell \Lambda_j^m \Lambda_k^n T^{i,j,k},$$

where Λ_v^u denotes the component of Λ in row u and in column v. The computations yield

$$\widetilde{T}^{1,1,1} = 0, \qquad \widetilde{T}^{2,1,1} = -1,$$
$$\widetilde{T}^{1,1,2} = 4, \qquad \widetilde{T}^{2,1,2} = -6,$$
$$\widetilde{T}^{1,2,1} = 1, \qquad \widetilde{T}^{2,2,1} = 0,$$
$$\widetilde{T}^{1,2,2} = -8, \qquad \widetilde{T}^{2,2,2} = 12.$$

EXERCISE 5.11. There are two conditions to check (cf. Sect. 2.1.2):

(1) The zero covariant k-tensor (associating the number zero to each k-tuple of vectors from V) is trivially symmetric, hence it is in $S^k V^*$.
(2) The subset $S^k V^*$ is closed under linear combinations. Assuming that T and U are symmetric covariant k-tensors, and $c, d \in \mathbb{R}$, we check that $cT + dU$ is symmetric, that is, independent of the order of the k arguments from V. This follows from the fact that the value of $cT + dU$ on vectors $v_1, \ldots, v_k \in V$ is the corresponding linear combination of the values $T(v_1, \ldots, v_k)$ and $U(v_1, \ldots, v_k)$, which by assumption do not depend on the order of the vectors.

The argument is similar for the set of all antisymmetric covariant k-tensors, $\bigwedge^k V^*$.

EXERCISE 5.12. Let $\beta^1, \ldots, \beta^n$ form a basis of V^* dual to a basis $v_1, \ldots, v_n$ of V. Each tensor $\frac{1}{2} \left(\beta^i \otimes \beta^j + \beta^j \otimes \beta^i \right)$ is symmetric by construction. In order to show that the tensors

$$\tfrac{1}{2} \left(\beta^i \otimes \beta^j + \beta^j \otimes \beta^i \right) \text{ with } i \geq j$$

form a basis of $S^2 V^*$, we check that (cf. Definition 2.22):

(1) they are *linearly independent* and
(2) they *span V*.

Suppose we have a linear combination (with coefficients $\alpha_{ij} \in \mathbb{R}$) yielding zero:

$$\sum_{i \geq j} \alpha_{ij} \tfrac{1}{2} \left(\beta^i \otimes \beta^j + \beta^j \otimes \beta^i \right) = 0 .$$

Evaluating this at a pair of basis vectors v_k, v_ℓ we get $\frac{1}{2} \alpha_{k\ell} = 0$ when $k > \ell$ and we get $\alpha_{kk} = 0$ when $k = \ell$, hence we conclude that it must be the trivial combination with all coefficients zero. This shows (1).

Now we show that any element $T \in S^2 V^*$ is a linear combination of the above elements. We define the numbers

$$\alpha_{ij} = \begin{cases} 2T(v_i, v_j) & \text{if } i > j \\ T(v_i, v_i) & \text{if } i = j. \end{cases}$$

We can conclude the following tensor equality

$$T = \sum_{i \geq j} \alpha_{ij} \tfrac{1}{2} \left(\beta^i \otimes \beta^j + \beta^j \otimes \beta^i \right)$$

from the fact that the difference of these two tensors vanishes on each pair of basis vectors v_k, v_ℓ, hence (2) holds.

To obtain the dimension formula, we simply count the number of elements in the basis set

$$\left\{ \tfrac{1}{2} \left(\beta^i \otimes \beta^j + \beta^j \otimes \beta^i \right) \mid i \geq j \right\}.$$

The index i can take all values $1, \ldots, n$. For each chosen i, the index j can take values $1, \ldots, i$, which is a number i of values, so we get the count

$$\sum_{i=1}^{n} i = \frac{n(n+1)}{2}$$

denoted $\binom{n}{2}$ in combinatorics.

The argument is similar for the case of $\bigwedge^2 V^*$.

EXERCISE 5.15. The coordinates of $b_1 \otimes a_1 + b_2 \otimes a_2$ with respect to the basis $b_i \otimes a_j$ are δ^{ij}. So the task amounts to showing that there is no solution to the equation

$$\begin{pmatrix} v^1 \\ v^2 \end{pmatrix} \begin{pmatrix} w^1 & w^2 \end{pmatrix} = \begin{pmatrix} 1 & 0 \\ 0 & 1 \end{pmatrix}$$

where the v^i and w^j are the coordinates of arbitrary vectors $v \in V$ and $w \in W$ w.r.t. the given bases. Indeed, the system

$$\begin{cases} v^1 w^1 = 1 \\ v^1 w^2 = 0 \\ v^2 w^1 = 0 \\ v^2 w^2 = 1 \end{cases}$$

has no solution, since the first two equations force that w^2 be zero, yet this is impossible for the last equation.

7 Solutions to Exercises

Chapter 6

EXERCISE 6.4. We choose a coordinate system with origin at the vertex O, with x-axis along the side of length a, y-axis along the side of length b and z-axis perpendicular to the plate. We already know that the mass density is constant equal to $\rho = \frac{m}{ab}$. Then

$$\underbrace{I_{11}}_{I_{xx}} = \int_0^a \int_0^b (y^2 + \underbrace{z^2}_{=0}) \underbrace{\rho}_{\frac{m}{ab}} \, dy \, dx$$

$$= \frac{m}{ab} a \int_0^b y^2 \, dy$$

$$= \frac{m}{b} \left[\frac{y^3}{3} \right]_0^b = \frac{m}{3} b^2.$$

Similarly,

$$\underbrace{I_{22}}_{I_{yy}} = \frac{m}{3} a^2,$$

and

$$\underbrace{I_{33}}_{I_{zz}} = \int_0^a \int_0^b (x^2 + y^2) \rho \, dy \, dx = \frac{m}{3}(a^2 + b^2)$$

is again just the sum of I_{11} and I_{22}.
Furthermore,

$$I_{23} = I_{32} = -\int_0^a \int_0^b y \underbrace{z}_{=0} \rho \, dy \, dx = 0,$$

and, similarly, $I_{31} = I_{13} = 0$. Finally, we have

$$I_{21} = I_{21} = -\int_0^a \int_0^b xy \rho \, dy \, dx = -\frac{m}{ab} \left[\frac{x^2}{2} \right]_0^a \left[\frac{y^2}{2} \right]_0^b = -\frac{mab}{4}.$$

We conclude that the inertia tensor is given by the matrix

$$\frac{m}{12} \begin{pmatrix} 4b^2 & -3ab & 0 \\ -3ab & 4a^2 & 0 \\ 0 & 0 & 4(a^2 + b^2) \end{pmatrix}.$$

EXERCISE 6.6. We again choose the unit vector

$$u = \begin{pmatrix} \frac{a}{\sqrt{a^2+b^2}} \\ \frac{b}{\sqrt{a^2+b^2}} \\ 0 \end{pmatrix} = \frac{1}{\sqrt{a^2+b^2}} \begin{pmatrix} a \\ b \\ 0 \end{pmatrix}$$

defining the axis of rotation, but now use the inertia tensor computed in Exercise 6.4,

$$\frac{m}{12} \begin{pmatrix} 4b^2 & -3ab & 0 \\ -3ab & 4a^2 & 0 \\ 0 & 0 & 4(a^2+b^2) \end{pmatrix},$$

thus obtaining

$$I_u = I_{ij} u^i u^j$$

$$= \frac{m}{12(a^2+b^2)} (a\ b\ 0) \begin{pmatrix} 4b^2 & -3ab & 0 \\ -3ab & 4a^2 & 0 \\ 0 & 0 & 4(a^2+b^2) \end{pmatrix} \begin{pmatrix} a \\ b \\ 0 \end{pmatrix}$$

$$= \frac{m}{6} \frac{a^2 b^2}{a^2+b^2},$$

necessarily the same result as in Example 6.5.

EXERCISE 6.7.

(1) We use the inertia tensor calculated in Example 6.3, where the origin of the coordinate system is at the center of mass, and choose $u = e_3$. The moment of inertia is then

$$I_u = I_{ij} u^i u^j$$

$$= \frac{m}{12} (0\ 0\ 1) \begin{pmatrix} b^2 & 0 & 0 \\ 0 & a^2 & 0 \\ 0 & 0 & a^2+b^2 \end{pmatrix} \begin{pmatrix} 0 \\ 0 \\ 1 \end{pmatrix}$$

$$= I_{33} = \frac{m}{12}(a^2+b^2).$$

(2) We use the inertia tensor calculated in Exercise 6.4, where the origin of the coordinate system is at a vertex of the plate, and choose $u = e_3$. The moment of inertia is then

$$I_u = I_{ij} u^i u^j$$

$$= \tfrac{m}{12} (0\ 0\ 1) \begin{pmatrix} 4b^2 & -3ab & 0 \\ -3ab & 4a^2 & 0 \\ 0 & 0 & 4(a^2+b^2) \end{pmatrix} \begin{pmatrix} 0 \\ 0 \\ 1 \end{pmatrix}$$

$$= I_{33} = \tfrac{m}{3}(a^2 + b^2).$$

EXERCISE 6.10.

(1) We choose the vector e_3 along the line L and complete it to an orthonormal basis for the vector space with origin at the center of mass. Since all atoms lie on the e_3-axis, they have $x_r^1 = x_r^2 = 0$, where the x_r^i denote the coordinates of particle r with respect to the basis above. Then we have:

$$I_{11} = \sum_{r=1}^{n} x_r^3 x_r^3 m_r = I_{22} =: I \quad \text{and} \quad I_{33} = 0 = I_{ij}, \text{ for } i \neq j,$$

hence, the inertia tensor is

$$(I_{ij}) = I \begin{pmatrix} 1 & 0 & 0 \\ 0 & 1 & 0 \\ 0 & 0 & 0 \end{pmatrix}.$$

The principal axes of inertia are the axis along L, corresponding to the principal moment 0, and the axes perpendicular to L, corresponding to the principal moment of inertia I. It remains to determine I. Denoting the total mass by $m := \sum_{r=1}^{n} m_r$, by the choice of origin the coordinates satisfy

$$x_r^3 = \frac{1}{m} \sum_{s=1}^{n} \left(x_r^3 - x_s^3 \right) m_s,$$

therefore,

$$I = \sum_{r=1}^{n} x_r^3 x_r^3 m_r = \frac{1}{m^2} \sum_{r,s,t=1}^{n} \underbrace{\left(x_r^3 - x_s^3\right)\left(x_r^3 - x_t^3\right)}_{(x_r^3-x_s^3)-(x_t^3-x_s^3)} m_r m_s m_t$$

$$= \frac{1}{m^2} \sum_{r,s,t=1}^{n} \underbrace{\left(x_r^3 - x_s^3\right)^2}_{\ell_{rs}^2} m_r m_s m_t$$

$$\underbrace{- \frac{1}{m^2} \sum_{r,s,t=1}^{n} \left(x_r^3 - x_s^3\right)\left(x_t^3 - x_s^3\right) m_r m_s m_t}_{=I} \;.$$

We conclude that

$$I = \frac{1}{2m} \sum_{r \neq s} \ell_{rs}^2 m_r m_s \;.$$

The ellipsoid of inertia has equation

$$I\left(x^1 x^1 + x^2 x^2\right) = 1 \;.$$

(2) Let e_1 be parallel to AB and e_3 perpendicular to the triangle. We complete the orthonormal basis with e_2 along the triangle's height. Let x_A^i, x_B^i, x_C^i be the coordinate vectors of atoms A, B, C with respect to the basis above. These vectors satisfy:

$$(x_B^i) = -\frac{1}{2m_1+m_2} \begin{pmatrix} am_1 + \frac{a}{2}m_2 \\ hm_2 \\ 0 \end{pmatrix}$$

$$(x_C^i) = \begin{pmatrix} a \\ 0 \\ 0 \end{pmatrix} - \frac{1}{2m_1+m_2} \begin{pmatrix} am_1 + \frac{a}{2}m_2 \\ hm_2 \\ 0 \end{pmatrix} = \frac{1}{2m_1+m_2} \begin{pmatrix} am_1 + \frac{a}{2}m_2 \\ -hm_2 \\ 0 \end{pmatrix}$$

$$(x_A^i) = \begin{pmatrix} \frac{a}{2} \\ h \\ 0 \end{pmatrix} - \frac{1}{2m_1+m_2} \begin{pmatrix} am_1 + \frac{a}{2}m_2 \\ hm_2 \\ 0 \end{pmatrix} = \frac{1}{2m_1+m_2} \begin{pmatrix} 0 \\ 2m_1 h \\ 0 \end{pmatrix}$$

Moreover, since $x_A^3 = x_B^3 = x_C^3 = 0$, we have:

$I_{11} = x_B^2 x_B^2 m_1 + x_C^2 x_C^2 m_1 + x_A^2 x_A^2 m_2 = \frac{2m_1 m_2 h^2}{2m_1 + m_2}$

$I_{22} = x_B^1 x_B^1 m_1 + x_C^1 x_C^1 m_1 + x_A^1 x_A^1 m_2 = \frac{m_1 a^2}{2}$

$I_{33} = x_B^1 x_B^1 m_1 + x_B^2 x_B^2 m_1 + x_C^1 x_C^1 m_1 + x_C^2 x_C^2 m_1$
$\quad + x_A^1 x_A^1 m_2 + x_A^2 x_A^2 m_2 = I_{11} + I_{22}$

$I_{13} = I_{23} = 0,\quad$ since we always multiply with terms with $x_*^3 = 0$

$I_{12} = \underbrace{(x_B^1 x_B^2 + x_C^1 x_C^2)}_{=0} m_1 + x_A^1 x_A^2 m_2 = 0$

The principal moments of inertia are I_{11}, I_{22} and I_{33}, and the coordinate axes are the corresponding principal axes of inertia. The ellipsoid of inertia has equation

$$\underbrace{\frac{2m_1 m_2 h^2}{2m_1 + m_2}}_{I_{11}} \left(x^1 x^1 + x^3 x^3\right) + \underbrace{\frac{m_1 a^2}{2}}_{I_{22}} \left(x^2 x^2 + x^3 x^3\right) = 1 \;.$$

(3) The tetrahedron consists of an equilateral triangle (a special case of the previous part) together with an extra point x_D. We choose the basis as above, so parts of the computation may be imported from the previous part with height $h = \frac{a}{2}\sqrt{3}$. The new coordinates are then slightly vertically shifted:

$$(x_B^i) = \frac{a}{2}\begin{pmatrix} -1 \\ -\frac{1}{\sqrt{3}} \\ -\frac{1}{\sqrt{6}} \end{pmatrix},\qquad (x_C^i) = \frac{a}{2}\begin{pmatrix} 1 \\ -\frac{1}{\sqrt{3}} \\ -\frac{1}{\sqrt{6}} \end{pmatrix},$$

$$(x_A^i) = \frac{a}{2}\begin{pmatrix} 0 \\ \frac{2}{\sqrt{3}} \\ -\frac{1}{\sqrt{6}} \end{pmatrix},\qquad (x_D^i) = \frac{a}{2}\begin{pmatrix} 0 \\ 0 \\ \frac{\sqrt{3}}{\sqrt{2}} \end{pmatrix}.$$

A computation as in the previous part yields

$$I_{ij} = ma^2 \delta_{ij} \;.$$

We have then a spherical top wih all principal moments of inertia equal to ma^2 and for which any axis (through the origin) is a principal axis of inertia. The equation of the ellipsoid of inertia is

$$ma^2 \left(x^1 x^1 + x^2 x^2 + x^3 x^3 \right) = 1 \,.$$

EXERCISE 6.11. We choose a positively oriented orthonormal basis along the sides of the parallelepiped. Then we compute:

$$I_{11} = \frac{m}{abc} \int_{-\frac{a}{2}}^{\frac{a}{2}} \int_{-\frac{b}{2}}^{\frac{b}{2}} \int_{-\frac{c}{2}}^{\frac{c}{2}} x^2 x^2 + x^3 x^3 \, dx^1 dx^2 dx^3 = \frac{m(b^2 + c^2)}{12}$$

$$I_{22} = \frac{m(a^2 + c^2)}{12}$$

$$I_{33} = \frac{m(a^2 + b^2)}{12}$$

$$I_{ij} = 0, \quad \text{when } i \neq j \,.$$

(1) Since the matrix (I_{ij}) is diagonal, the corresponding principal moments of inertia are its diagonal entries I_{11}, I_{22}, I_{33} and the corresponding principal axes of inertia are the coordinate axes (parallel to the sides of the parallelepiped).
(2) The equation of the ellipsoid of inertia is

$$m((b^2 + c^2)x^1 x^1 + (a^2 + c^2)x^2 x^2 + (a^2 + b^2)x^3 x^3) = 12 \,.$$

(3) The kinetic energy of K is

$$E = \frac{1}{2} I_{ij} \omega^i \omega^j \,.$$

(4) The (covariant) coordinates of the angular momentum of K are

$$L_i = I_{ij} w^j \,.$$

7 Solutions to Exercises

EXERCISE 6.20.

(1) The principal stresses are the eigenvalues of σ, that is, the roots of the corresponding characteristic polynomial:

$$p_\sigma(\lambda) = \det(\sigma - \lambda \operatorname{Id}) = \det \begin{bmatrix} -\lambda & 0 & 6 \\ 0 & 1-\lambda & 0 \\ 6 & 0 & 5-\lambda \end{bmatrix}$$

$$= -\lambda(1-\lambda)(5-\lambda) - 36(1-\lambda) = (1-\lambda)(\lambda^2 - 5\lambda - 36).$$

The quadratic factor $\lambda^2 - 5\lambda - 36$ has the roots

$$\frac{5 \pm \sqrt{25 + 4 \cdot 36}}{2} = \frac{5 \pm \sqrt{169}}{2} = \frac{5 \pm 13}{2} = 9 \text{ or } -4.$$

Therefore, the principal stresses of σ are (up to reordering)

$$\sigma^1 := -4, \qquad \sigma^2 := 1 \quad \text{and} \quad \sigma^3 := 9.$$

(2) The principal directions are defined by corresponding eigenvectors v_1, v_2, v_3, which we obtain by finding non-trivial solutions of

$$(\sigma + 4\operatorname{Id})v_1 = 0$$
$$(\sigma - \operatorname{Id})v_2 = 0$$
$$(\sigma - 9\operatorname{Id})v_3 = 0,$$

or, equivalently,

$$\begin{pmatrix} 4 & 0 & 6 \\ 0 & 5 & 0 \\ 6 & 0 & 9 \end{pmatrix} v_1 = 0$$

$$\begin{pmatrix} -1 & 0 & 6 \\ 0 & 0 & 0 \\ 6 & 0 & 4 \end{pmatrix} v_2 = 0$$

$$\begin{pmatrix} -9 & 0 & 6 \\ 0 & -8 & 0 \\ 6 & 0 & -4 \end{pmatrix} v_3 = 0.$$

We choose

$$v_1 = \begin{pmatrix} 3 \\ 0 \\ -2 \end{pmatrix}, v_2 = \begin{pmatrix} 0 \\ 1 \\ 0 \end{pmatrix}, \text{ and } v_3 = \begin{pmatrix} 2 \\ 0 \\ 3 \end{pmatrix}$$

to span the principal directions.

(3) The basis $\mathcal{B}$ will be an orthonormal eigenbasis of σ and the corresponding diagonal matrix will have the corresponding eigenvalues of σ along the diagonal. We simply normalize the above eigenvectors, since they are already orthogonal (because they correspond to different eigenvalues):

$$b_1 = \frac{v_1}{\|v_1\|} = \frac{1}{\sqrt{13}} \begin{bmatrix} 3 \\ 0 \\ -2 \end{bmatrix},$$

$$b_2 = \frac{v_2}{\|v_2\|} = \begin{bmatrix} 0 \\ 1 \\ 0 \end{bmatrix} \text{ and } b_3 = \frac{v_3}{\|v_3\|} = \frac{1}{\sqrt{13}} \begin{bmatrix} 2 \\ 0 \\ 3 \end{bmatrix}.$$

With respect to the basis $\mathcal{B} = \{b_1, b_2, b_3\}$, the matrix representation of σ is

$$\begin{bmatrix} \sigma^1 & 0 & 0 \\ 0 & \sigma^2 & 0 \\ 0 & 0 & \sigma^3 \end{bmatrix} = \begin{bmatrix} -4 & 0 & 0 \\ 0 & 1 & 0 \\ 0 & 0 & 9 \end{bmatrix}.$$

(4) The stress invariants I_1, I_2 and I_3 can be obtained from the principal stresses (or eigenvalues) of σ by:

$$I_1 = \text{tr}\,\sigma = \sigma^1 + \sigma^2 + \sigma^3 = -4 + 1 + 9 = 6$$
$$I_2 = -\sigma^1\sigma^2 - \sigma^2\sigma^3 - \sigma^3\sigma^1 = -(-4 + 9 - 36) = 31$$
$$I_3 = \det(\sigma) = \sigma^1\sigma^2\sigma^3 = -4 \cdot 1 \cdot 9 = -36$$

Alternatively, we may take the coefficients of the characteristic polynomial $p_\sigma(\lambda)$ and adjust the sign.

7 Solutions to Exercises

EXERCISE 6.22.

(1) The given stress tensor has trace

$$\text{tr}(\sigma) = \sigma^{11} + \sigma^{22} + \sigma^{33} = -2 + 2 + 6 = 6.$$

We hence define

$$\sigma_S = \sigma - \tfrac{1}{3}\text{tr}(\sigma)\text{Id} = \begin{bmatrix} -2 & 0 & 3 \\ 0 & 2 & 0 \\ 3 & 0 & 6 \end{bmatrix} - \begin{bmatrix} 2 & 0 & 0 \\ 0 & 2 & 0 \\ 0 & 0 & 2 \end{bmatrix} = \begin{bmatrix} -4 & 0 & 3 \\ 0 & 0 & 0 \\ 3 & 0 & 4 \end{bmatrix}$$

and $\sigma_P = \tfrac{1}{3}\text{tr}(\sigma)\text{Id} = \begin{bmatrix} 2 & 0 & 0 \\ 0 & 2 & 0 \\ 0 & 0 & 2 \end{bmatrix}$.

(2) The diagonal matrix D will have the eigenvalues of σ_S along the diagonal and the seeked orthonormal basis is an orthonormal eigenbasis of σ_S. We, thus, first determine the eigenvalues of σ_S:

$$p_{\sigma_S}(\lambda) = \det(\sigma_S - \lambda\text{Id})$$

$$= \det \begin{bmatrix} -4-\lambda & 0 & 3 \\ 0 & -\lambda & 0 \\ 3 & 0 & 4-\lambda \end{bmatrix} = -\lambda \cdot \det \begin{bmatrix} -4-\lambda & 3 \\ 3 & 4-\lambda \end{bmatrix}$$

$$= -\lambda(\lambda^2 - 16 - 9) = -\lambda(\lambda - 5)(\lambda + 5).$$

This factorization of the characteristic polynomial shows that the eigenvalues of σ_S are $\lambda_1 = -5$, $\lambda_2 = 0$ and $\lambda_3 = 5$. Next, we determine corresponding orthonormal eigenvectors v_1, v_2, v_3 of σ_S (with respect to the basis $\mathcal{E}$):

$$E_{-5} = \ker(\sigma_S + 5\text{Id}) = \ker \begin{bmatrix} 1 & 0 & 3 \\ 0 & 5 & 0 \\ 3 & 0 & 9 \end{bmatrix} = \left\{ \begin{bmatrix} 3x \\ 0 \\ -x \end{bmatrix} : x \in \mathbb{R} \right\}$$

$$E_0 = \ker(\sigma_S) = \ker \begin{bmatrix} -4 & 0 & 3 \\ 0 & 0 & 0 \\ 3 & 0 & 4 \end{bmatrix} = \left\{ \begin{bmatrix} 0 \\ x \\ 0 \end{bmatrix} : x \in \mathbb{R} \right\}$$

$$E_5 = \ker(\sigma_S - 5\text{Id}) = \ker \begin{bmatrix} -9 & 0 & 3 \\ 0 & -5 & 0 \\ 3 & 0 & -1 \end{bmatrix} = \left\{ \begin{bmatrix} x \\ 0 \\ 3x \end{bmatrix} : x \in \mathbb{R} \right\}.$$

We choose $x = 1$ to obtain concrete vectors, which we normalize to define

$$[v_1]_{\mathcal{E}} = \frac{1}{\sqrt{10}} \begin{bmatrix} 3 \\ 0 \\ -1 \end{bmatrix}, \quad [v_2]_{\mathcal{E}} = \begin{bmatrix} 0 \\ 1 \\ 0 \end{bmatrix} \quad \text{and} \quad [v_3]_{\mathcal{E}} = \frac{1}{\sqrt{10}} \begin{bmatrix} 1 \\ 0 \\ 3 \end{bmatrix}.$$

The three vectors above form an orthonormal eigenbasis, $\mathcal{D}$, for the deviatoric stress σ_S. The change of basis matrix, L, from $\mathcal{E}$ to $\mathcal{D}$ has columns given by the components of the above $[v_1]_{\mathcal{E}}, [v_2]_{\mathcal{E}}, [v_3]_{\mathcal{E}}$. With respect to the basis $\mathcal{D}$, the matrix representation of the deviatoric stress becomes diagonal:

$$[\sigma_S]_{\mathcal{D}} = \underbrace{\begin{bmatrix} \frac{3}{\sqrt{10}} & 0 & \frac{-1}{\sqrt{10}} \\ 0 & 1 & 0 \\ \frac{1}{\sqrt{10}} & 0 & \frac{3}{\sqrt{10}} \end{bmatrix}}_{L^{-1} = {}^t L} \underbrace{\begin{bmatrix} -4 & 0 & 3 \\ 0 & 0 & 0 \\ 3 & 0 & 4 \end{bmatrix}}_{\sigma_S} \underbrace{\begin{bmatrix} \frac{3}{\sqrt{10}} & 0 & \frac{1}{\sqrt{10}} \\ 0 & 1 & 0 \\ \frac{-1}{\sqrt{10}} & 0 & \frac{3}{\sqrt{10}} \end{bmatrix}}_{L} = \underbrace{\begin{bmatrix} -5 & 0 & 0 \\ 0 & 0 & 0 \\ 0 & 0 & 5 \end{bmatrix}}_{\mathcal{D}}.$$

(3) When symmetric matrices have the same eigenvalues, we can diagonalize each of them to the same diagonal matrix (containing the eigenvalues along the diagonal) via orthogonal changes of basis relying on their eigenvectors. In part (2), we obtained this diagonalization for the matrix of σ_S with respect to $\mathcal{E}$. Now we seek a matrix of the form

$$A = \begin{bmatrix} 0 & x & y \\ x & 0 & z \\ y & z & 0 \end{bmatrix}$$

having the same eigenvalues as σ_S. The characteristic polynomial of such A,

$$p_A(\lambda) = \det(A - \lambda \mathrm{Id}) = \begin{bmatrix} -\lambda & x & y \\ x & -\lambda & z \\ y & z & -\lambda \end{bmatrix} = -\lambda^3 + 2xyz + (x^2 + y^2 + z^2)\lambda,$$

must then have the zeros $-5, 0, 5$, that is

$$\begin{cases} 0 + 2xyz + 0 = 0 \\ 125 + 2xyz - 5(x^2 + y^2 + z^2) = 0 \\ -125 + 2xyz + 5(x^2 + y^2 + z^2) = 0 \end{cases} \iff \begin{cases} xyz = 0 \\ x^2 + y^2 + z^2 = 25 \end{cases}$$

For instance, the triple $x = z = 0$ and $y = 5$ satisfies the above conditions. We thus choose the matrix

$$A = \begin{bmatrix} 0 & 0 & 5 \\ 0 & 0 & 0 \\ 5 & 0 & 0 \end{bmatrix}.$$

Next, we determine corresponding orthonormal eigenvectors w_1, w_2, w_3 of A (with respect to the basis $\mathcal{E}$):

$$\ker(A + 5\mathrm{Id}) = \ker \begin{bmatrix} 5 & 0 & 5 \\ 0 & 5 & 0 \\ 5 & 0 & 5 \end{bmatrix} = \left\{ \begin{bmatrix} x \\ 0 \\ -x \end{bmatrix} : x \in \mathbb{R} \right\}$$

$$\ker(A) = \ker \begin{bmatrix} 0 & 0 & 5 \\ 0 & 0 & 0 \\ 5 & 0 & 0 \end{bmatrix} = \left\{ \begin{bmatrix} 0 \\ x \\ 0 \end{bmatrix} : x \in \mathbb{R} \right\}$$

$$\ker(A - 5\mathrm{Id}) = \ker \begin{bmatrix} -5 & 0 & 5 \\ 0 & -5 & 0 \\ 5 & 0 & -5 \end{bmatrix} = \left\{ \begin{bmatrix} x \\ 0 \\ x \end{bmatrix} : x \in \mathbb{R} \right\}.$$

We choose $x = 1$ to obtain concrete vectors and normalize them to define

$$[w_1]_{\mathcal{E}} = \frac{1}{\sqrt{2}} \begin{bmatrix} 1 \\ 0 \\ -1 \end{bmatrix}, \quad [w_2]_{\mathcal{E}} = \begin{bmatrix} 0 \\ 1 \\ 0 \end{bmatrix} \quad \text{and} \quad [w_3]_{\mathcal{E}} = \frac{1}{\sqrt{2}} \begin{bmatrix} 1 \\ 0 \\ 1 \end{bmatrix}.$$

The three vectors above form an orthonormal eigenbasis, $\mathcal{D}'$, for A. Just as in part (2), we obtain the diagonalization:

$$[A]_{\mathcal{D}'} = \underbrace{\begin{bmatrix} \frac{1}{\sqrt{2}} & 0 & \frac{-1}{\sqrt{2}} \\ 0 & 1 & 0 \\ \frac{1}{\sqrt{2}} & 0 & \frac{1}{\sqrt{2}} \end{bmatrix}}_{M^{-1} = {}^t M} \underbrace{\begin{bmatrix} 0 & 0 & 5 \\ 0 & 0 & 0 \\ 5 & 0 & 0 \end{bmatrix}}_{A} \underbrace{\begin{bmatrix} \frac{1}{\sqrt{2}} & 0 & \frac{1}{\sqrt{2}} \\ 0 & 1 & 0 \\ \frac{-1}{\sqrt{2}} & 0 & \frac{1}{\sqrt{2}} \end{bmatrix}}_{M} = \underbrace{\begin{bmatrix} -5 & 0 & 0 \\ 0 & 0 & 0 \\ 0 & 0 & 5 \end{bmatrix}}_{D}.$$

Putting the two equations (for $[\sigma_S]_{\mathcal{D}}$ and for $[A]_{\mathcal{D}'}$) together, we obtain

$$L^{-1} \sigma_S L = D = M^{-1} A M$$

from which we find

$$A = M L^{-1} \sigma_S L M^{-1}.$$

So, the change of basis matrix is

$$L_{\mathcal{E}\mathcal{B}} = LM^{-1} = \begin{bmatrix} \frac{3}{\sqrt{10}} & 0 & \frac{1}{\sqrt{10}} \\ 0 & 1 & 0 \\ \frac{-1}{\sqrt{10}} & 0 & \frac{3}{\sqrt{10}} \end{bmatrix} \begin{bmatrix} \frac{1}{\sqrt{2}} & 0 & \frac{-1}{\sqrt{2}} \\ 0 & 1 & 0 \\ \frac{1}{\sqrt{2}} & 0 & \frac{1}{\sqrt{2}} \end{bmatrix} = \begin{bmatrix} \frac{2}{\sqrt{5}} & 0 & \frac{-1}{\sqrt{5}} \\ 0 & 1 & 0 \\ \frac{1}{\sqrt{5}} & 0 & \frac{2}{\sqrt{5}} \end{bmatrix}$$

and the desired basis $\mathcal{B}$ is formed by vectors with

$$[b_1]_{\mathcal{E}} = \frac{1}{\sqrt{5}} \begin{bmatrix} 2 \\ 0 \\ 1 \end{bmatrix}, \quad [b_2]_{\mathcal{E}} = \begin{bmatrix} 0 \\ 1 \\ 0 \end{bmatrix} \quad \text{and} \quad [b_3]_{\mathcal{E}} = \frac{1}{\sqrt{5}} \begin{bmatrix} -1 \\ 0 \\ 2 \end{bmatrix}.$$

Reality check: We compute $[\sigma_S]_{\mathcal{B}}$ directly

$$[\sigma_S]_{\mathcal{B}} = L_{\mathcal{E}\mathcal{B}}^{-1} [\sigma_S]_{\mathcal{E}} L_{\mathcal{E}\mathcal{B}} = {}^tL_{\mathcal{E}\mathcal{B}} [\sigma_S]_{\mathcal{E}} L_{\mathcal{E}\mathcal{B}}$$

$$= \begin{bmatrix} \frac{2}{\sqrt{5}} & 0 & \frac{1}{\sqrt{5}} \\ 0 & 1 & 0 \\ -\frac{1}{\sqrt{5}} & 0 & \frac{2}{\sqrt{5}} \end{bmatrix} \begin{bmatrix} -4 & 0 & 3 \\ 0 & 0 & 0 \\ 3 & 0 & 4 \end{bmatrix} \begin{bmatrix} \frac{2}{\sqrt{5}} & 0 & -\frac{1}{\sqrt{5}} \\ 0 & 1 & 0 \\ \frac{1}{\sqrt{5}} & 0 & \frac{2}{\sqrt{5}} \end{bmatrix}$$

$$= \begin{bmatrix} -\sqrt{5} & 0 & 2\sqrt{5} \\ 0 & 0 & 0 \\ 2\sqrt{5} & 0 & \sqrt{5} \end{bmatrix} \begin{bmatrix} \frac{2}{\sqrt{5}} & 0 & -\frac{1}{\sqrt{5}} \\ 0 & 1 & 0 \\ \frac{1}{\sqrt{5}} & 0 & \frac{2}{\sqrt{5}} \end{bmatrix} = \begin{bmatrix} 0 & 0 & 5 \\ 0 & 0 & 0 \\ 5 & 0 & 0 \end{bmatrix}.$$

EXERCISE 6.26. We follow Sect. 6.2.5. Just like stress, the strain tensor is symmetric

$$\mathcal{E} = \begin{pmatrix} \varepsilon_{11} & \varepsilon_{12} & \varepsilon_{13} \\ \varepsilon_{12} & \varepsilon_{22} & \varepsilon_{23} \\ \varepsilon_{13} & \varepsilon_{23} & \varepsilon_{33} \end{pmatrix}.$$

Let $k := \operatorname{tr}\varepsilon/3 = (\varepsilon^{11} + \varepsilon^{22} + \varepsilon^{33})/3$ and define the *uniform compression*

$$\begin{pmatrix} k & 0 & 0 \\ 0 & k & 0 \\ 0 & 0 & k \end{pmatrix}$$

7 Solutions to Exercises

as well as the traceless difference

$$s^{ij} := \varepsilon^{ij} - k\delta^{ij} = \begin{pmatrix} \varepsilon^{11} - k & \varepsilon^{12} & \varepsilon^{13} \\ \varepsilon^{12} & \varepsilon^{22} - k & \varepsilon^{23} \\ \varepsilon^{13} & \varepsilon^{23} & \varepsilon^{33} - k \end{pmatrix},$$

so that, we have

$$\varepsilon^{ij} = s^{ij} + k\delta^{ij}.$$

By Fact 6.21, the above traceless difference s^{ij} may be represented with respect to some orthonormal basis as a *shear deformation*, i.e., as a strain tensor of the form

$$\begin{pmatrix} 0 & \tilde{\varepsilon}^{12} & \tilde{\varepsilon}^{13} \\ \tilde{\varepsilon}^{12} & 0 & \tilde{\varepsilon}^{23} \\ \tilde{\varepsilon}^{13} & \tilde{\varepsilon}^{23} & 0 \end{pmatrix}.$$

Bibliography

1. Akivis, M., & Goldberg, V. (2003). *Tensor calculus with applications*. World Scientific.
2. Axler, S. (1996) *Linear algebra done right*. Springer.
3. Bretscher, O. (1997) *Linear algebra with applications*. Prentice Hall.
4. de Souza Sánchez Filho, E. (2016). *Tensor calculus for engineers and physicists*. Springer.
5. Dodson, C. T. J. (1991). *Tensor geometry: The geometric viewpoint and its uses. Graduate texts in mathematics* (2nd ed.). Springer.
6. Hass, J., Heil, C., & Weir, M. (2017). *Thomas' calculus: Early transcendentals* (14th ed.). Pearson.
7. Hess, S. (2015). *Tensors for physics. Undergraduate lecture notes in physics*. Springer.
8. Jeevanjee, N. (2015). *An introduction to tensors and group theory for physicists* (2nd ed.). Birkhäuser.
9. Landsberg, J. M. (2012). *Tensors: Geometry and applications*. American Mathematical Society.
10. Neuenschwander, D. (2015). *Tensor calculus for physics, A concise guide*. Johns Hopkins University Press.
11. Simmonds, J. (1994). *A brief on tensor analysis. Undergraduate texts in mathematics* (2nd ed.). Springer.
12. Stewart, J., Clegg, D., & Watson, S. (2020). *Calculus: Early transcendentals* (9th ed.). Cengage Learning.
13. Strang, G. (1980). *Linear algebra and its applications* (2nd ed.). Academic Press.
14. Tipler, P., & Mosca, G. (2008). *Physics for scientists and engineers* (6th ed.). W. H. Freeman and Company.
15. Walker, J. (2014). *Halliday and Resnick's principles of physics* (10th ed.). Wiley.

Index

A
Acceleration, 83
Angular momentum, 112
  total, 113
Anisotropic, 135
Antisymmetric, 130

B
Basis, 1, 14
  – of Bil($V \times V, \mathbb{R}$), 50
  dual, 42
  dual –, 41
    contravariance of the – –, 45
  eigen–, 32
  orthonormal –, 61, 64
  reciprocal –, 71
    contravariance of the – –, 75
    properties of the – –, 73
  standard –, 14
Boltzmann Axiom, 121

C
Canonical identification, 80
Canonical isomorphism, 87
Change of basis, 21
Characteristic polynomial, 30, 125
Component, 1
  normal –, 122
  shear –, 122, 132
Contravariance, vi, vii, 2, 3, 43, 46, 59, 77, 81, 84, 85, 88, 91, 93, 103, 126
Contravariant, 22, 45
Coordinate, 15
  covariant –s, 76
  – form, 38
  – vector, 15
Covariance, vi, vii, 2, 3, 42, 43, 46, 54, 59, 77, 81, 83–85, 88, 93, 107

  – of bilinear forms, 53
  – of linear forms, 42
Covariant, 43
  – tensor, 55
Covector, 37
Cross product, 47

D
Determinant, 31, 57
Dimension, 15
Displacement, 83
Dot product, 24, 47
Dual space, 39

E
Echelon form, 16
Eigen-
  –basis, 32
  –space, 32
  –value, 31, 32
  –vector, 32
Einstein convention, 18
Elasticity, 135
Electrical
  – conductivity, 135, 137
    – – tensor, 136
  – current density, 135
  – field, 135
  – resistivity tensor, 136
Energy, 83
  kinetic, 84, 104, 107, 115, 116, 164
  potential, 84
  total kinetic, 105, 106
Energytotal kinetic, 111
Euclidean space, 8
External exertion, 133
  material reaction to -, 133

F
Force
  attractive –, 117
  body –, 117, 119
  centrifugal –, 117
  electric, 135
  moment of –, 122
  surface –, 117, 119
Form
  bilinear –, 47, 49, 59, 92
    covariance of – –, 53
    positive definite – –, 59
    – symmetric –, 59
  coordinate –, 38
  linear –, 37, 49
    tensor product of – –s, 90
  multilinear – of order k, 54
  quadratic –, 62
    indefinite – –, 62
    negative definite – –, 62
    negative semidefinite – –, 62
    positive definite – –, 62
    positive semidefinite – –, 62
  trilinear –, 54
Fourier Heat Conduction Law, 137
Frame of reference, 1
Fundamental Theorem of Algebra, 31

G
Gauss–Jordan elimination, 15
Generalized Ohm Law, 136
General linear group, 21
Gradient, 83
Gram–Schmidt orthogonalization process, 67
Gravity, 117

H
Heat conductivity, 135
Heat flux, 137
Hooke's law, 133
Hydrostatic pressure, 125, 126

I
Image, 12
Index
  dummy, 25
  free, 19
Inertia
  ellipsoid of –, 115
  moments of –, 108, 110
    principal – –, 115

  polar – –, 108
  principal axes of –, 115
  products of –, 108
  total moment of –, 110
Inner product, 59
  components of the – –, 63
  standard –, 60
Isotropic crystal, 136

J
Jordan canonical form, 32

K
Kernel, 12
Kronecker delta symbol, 23

L
Lagrange identity, 106
Laplace expansion formula, 48
Legendre polynomials, 61
Linear
  – combination, 11
  – dependence, 13
  – functional, 37
  – independence, 12
  – space, 7
  – transformation, 10, 25

M
Magnetic
  – fluid density, 1
  – intensity, 1
Magnetization, 2
Map
  linear –, 1
  multilinear, 1
Material reaction, 133
Matrix
  augmented –, 16
  – of a bilinear form, 53
  – of a change of basis, 144
  – of the change of basis, 21
  conjugate –, 30
  diagonalizable –, 32
  identity –, 23
  – of a linear transformation, 25
  orthogonal –, 85
  standard –, 27
  symmetric –, 11, 61
    indefinite – –, 62

Index

negative definite – –, 62
negative semidefinite – –, 62
positive definite – –, 62
positive semidefinite – –, 62
Metric, 61
Momentum, 84

N
Norm, 60
Null space, 12

O
Order, 3

P
Parallelogram law, 8
Permeability
  scalar –, 2
  tensor –, 2
Permutation
  even –, 47, 56, 57
  odd –, 47, 56, 57
Position, 83
Precession, 115
Principal
  – coefficient, 132
  – direction, 123, 132
  – direction of heat conductivity, 138
  – planes, 123
  – stress, 123
  – coefficient, 138
Product
  cross –, 47
  dot –, 24, 47
  inner –, 59
    components of the –, 63
    standard – –, 60
  scalar –, 47
  scalar triple –, 47, 51, 55
  tensor –, 50, 57
    – – of linear forms, 49, 90
    – – of multilinear forms, 57
    – – of tensors, 96
    – – of vector spaces, 98
Projection
  orthogonal –, 28, 67
  parallel –, 81
Pseudovector, 103
Pure shear, 124

Q
Quadratic form, 62

R
Range, 12
Rank-Nullity Theorem, 88
Rate of electric current, 135
Rotation, 130

S
Scalar, 8
Scalar product, 47
Scalar triple product, 47, 51, 55
Shear
  – deformation, 124, 133
  pure –, 124
Span, 14
Spectral Theorem, 63, 114, 122, 127
Static equilibrium, 117
Strain, 131
  – energy density functional, 134
Stress, 117
  biaxial –, 123
  homogeneous –, 117
  – invariants, 125
  plane –, 123
  principal –, 123
  – tensor
    deviatoric – –, 126
    hydrostatic – –, 126
  uniaxial –, 123
Subscript, 18
Subspace, 11
Superscript, 18
Symmetries, 131
  major –, 134
  minor –, 133

T
Temperature, 137
Tensor, v, 3, 4, 26, 37, 43, 54, 55, 89, 90, 92, 97, 99
  antisymmetric –, 95
    (totally) – –, 95
  compliance –, 133
  components of a –, 93
  contravariance of –, 91
  contravariant –, 4, 92
  covariant –, 43, 54, 55, 83
  curvature –, v
  decomposable –, 97, 99

deformation –, 129
deviatoric stress –, 126
elasticity –, v, 133
electrical conductivity –, v, 136
electrical resistivity –, v, 136
electromagnetic –, v
elementary –, 97, 99
field –, 133
heat conductivity –, 137
heat resistivity –, 138
hydrostatic stress –, 126
inertia –, v, 107
magnetic susceptibility –, v
matter –, 133
order of a –, 93
permeability –, 2
– power, 101
– product
    – – of linear forms, 49, 90
    – – of multilinear forms, 57
    – – of tensors, 96
    – – of vector spaces, 98
– product of V^* and V^*, 50
pure –, 97, 99
purely contravariant –, 93
purely covariant –, 93
rank of a –, 97
rotation –, 129
simple –, 97, 99
skewsymmetric –
    (totally) – –, 95
stiffness –, 133
strain –, 129
stress –, v, 118
stress energy –, v
symmetric –, 95, 121
    (totally) – –, 95
thermal conductivity –, 137
zero –, 94

Top
  asymmetric –, 115
  spherical –, 115
  symmetric –, 115
Torque, 122
Trace, 31
Transposition, 10
Triangular slice, 118

U
Uniform compression, 133

V
Vector, v, 1, 8
  co–, 37
  coordinate –, 15
  length of a –, 60
  magnitude of a –, 60
  orthogonal –s, 60
  orthonormal –s, 60
  perpendicular –s, 60
  position –, 104
  pseudo–, 103
  unit –, 60
Vector space, 7
  bidual of a –, 87
  complex –, 32
  dual of a –, 39
  isomorphism of a –, 79
Velocity, 83
  angular, 103
  linear, 104

W
Work, 81

The manufacturer's authorised representative in the EU is Springer Nature Customer Service Centre GmbH, Europaplatz 3, 69115 Heidelberg, Germany. If you have any concerns regarding our products, please contact ProductSafety@springernature.com

Printed and bound by CPI Group (UK) Ltd, Croydon, CR0 4YY
26/03/2026
02078988-0002